高级大数据人才培养丛书

大数据实验手册

主　编　刘　鹏

副主编　叶晓江　朱光耀　杨震宇

電子工業出版社

Publishing House of Electronics Industry

北京•BEIJING

内 容 简 介

本书是中国大数据专家委员会刘鹏教授团队的心血之作。针对高校大数据相关专业实践教学以及个人提升大数据动手能力的需求，带领大数据研发团队，经过反复实践、提炼和验证形成本书。本书主要内容包括 HDFS 实验、YARN 实验、MapReduce 实验、Hive 实验、Spark 实验、ZooKeeper 实验、HBase 实验、Storm 实验、MongoDB 实验、LevelDB 实验、Mahout 实验和综合实战等。每个实验呈现详细的实验目的、实验内容、实验原理和实验流程。在线大数据实验平台（https://bd.cstor.cn）或 BDRack 大数据实验一体机可为全部实验提供完整的支撑。

"让学习变得轻松"是本书的初衷。本书填补了大数据教学过程的缺失环节，可培养学生实操动手和自主设计的能力。本书适合作为相关专业本科和研究生的实验手册，也可作为高职高专学校选做的实验教学内容，同时还可作为大数据从业人员的自学书籍。

图书在版编目（CIP）数据

大数据实验手册 / 刘鹏主编. —北京：电子工业出版社，2017.6
（高级大数据人才培养丛书）
ISBN 978-7-121-31618-0

Ⅰ. ①大… Ⅱ. ①刘… Ⅲ. ①数据处理—手册 Ⅳ. ①TP274-62

中国版本图书馆 CIP 数据核字（2017）第 107714 号

策划编辑：董亚峰
责任编辑：董亚峰　　　文字编辑：徐　烨
印　　刷：北京虎彩文化传播有限公司
装　　订：北京虎彩文化传播有限公司
出版发行：电子工业出版社
　　　　　北京市海淀区万寿路 173 信箱　邮编：100036
开　　本：787×1092　1/16　印张：18.25　字数：427 千字
版　　次：2017 年 6 月第 1 版
印　　次：2019 年 6 月第 5 次印刷
定　　价：45.00 元

凡所购买电子工业出版社图书有缺损问题，请向购买书店调换。若书店售缺，请与本社发行部联系，联系及邮购电话：（010）88254888，88258888。

质量投诉请发邮件至 zlts@phei.com.cn，盗版侵权举报请发邮件至 dbqq@phei.com.cn。

本书咨询联系方式：（010）88254694。

编 写 组

主　编：刘　鹏

副主编：叶晓江　朱光耀　杨震宇

编　委：戎新堃　苏泽月　吴荣荣

沈大为　谢　超　方龙双

武郑浩　朱纪光　张　燕

刘　明　孔　坎　石　胡

董广明

总　序

短短几年间，大数据就以一日千里的发展速度，快速实现了从概念到落地，直接带动了相关产业井喷式发展。全球多家研究机构统计数据显示，大数据产业将迎来发展黄金期：IDC 预计，大数据和分析市场将从 2016 年的 1300 亿美元增长到 2020 年的 2030 亿美元以上；中国报告大厅发布的大数据行业报告数据也说明，自 2017 年起,我国大数据产业将迎来发展黄金期，未来 2～3 年的市场规模增长率将保持在 35%左右。

数据采集、数据存储、数据挖掘、数据分析等大数据技术在越来越多的行业中得到应用，随之而来的就是大数据人才问题的凸显。麦肯锡预测，每年数据科学专业的应届毕业生将增加 7%，然而仅高质量项目对于专业数据科学家的需求每年就会增加 12%，完全供不应求。根据《人民日报》的报道，未来 3~5 年，中国需要 180 万数据人才，但目前只有约 30 万人，人才缺口达到 150 万之多。

以贵州大学为例，其首届大数据专业研究生就业率就达到 100%，可以说“一抢而空”。急切的人才需求直接催热了大数据专业，国家教育部正式设立“数据科学与大数据技术”本科新专业。目前已经有两批共计 35 所大学获批，包括北京大学、中南大学、对外经济贸易大学、中国人民大学、北京邮电大学、复旦大学等。估计 2018 年会有几百所高校获批。

不过，就目前而言，在大数据人才培养和大数据课程建设方面，大部分高校仍然处于起步阶段，需要探索的还有很多。首先，大数据是个新生事物，懂大数据的老师少之又少，院校缺“人”；其次，尚未形成完善的大数据人才培养和课程体系，院校缺“机制”；再次，大数据实验需要为每位学生提供集群计算机，院校缺“机器”；最后，院校没有海量数据，开展大数据教学科研工作缺“原材料”。

其实，早在网格计算和云计算兴起时，我国科技工作者就曾遇到过类似的挑战，我有幸参与了这些问题的解决过程。为了解决网格计算问题，我在清华大学读博期间，于 2001 年创办了中国网格信息中转站网站，每天花几个小时收集和分享有价值的资料给学术界，此后我也多次筹办和主持全国性的网格计算学术会议，进行信息传递与知识分享。2002 年，我与其他专家合作的《网格计算》教材也正式面世。

2008年，当云计算开始萌芽之时，我创办了中国云计算网站（chinacloud.cn）（在各大搜索引擎“云计算”关键词中排名第一），2010 年出版了《云计算（第一版）》、2011年出版了《云计算（第二版）》、2015 年出版了《云计算（第三版）》，每一版都花费了大量成本制作并免费分享对应的几十个教学 PPT。目前，这些 PPT 的下载总量达到了几百万次之多。同时，《云计算》教材也成为国内高校的首选教材，在 CNKI 公布的高被引图书名单中，对于 2010 年以来出版的所有图书，《云计算（第一版）》在自动化和计算机领域排名全国第一。除了资料分享，在 2010 年，我也在南京组织了全国高校云计算师资培训班，培养了国内第一批云计算老师，并通过与华为、中兴、360 等知名企业合作，输出云计算技术，培养云计算研发人才。这些工作获得了大家的认可与好评，此后我接连担任了工信部云计算研究中心专家、中国云计算专家委员会云存储组组长等职位。

近几年，面对日益突出的大数据发展难题，我也正在尝试使用此前类似的办法去应对这些挑战。为了解决大数据技术资料缺乏和交流不够通透的问题，我于 2013 年创办了中国大数据网站（thebigdata.cn），投入大量的人力进行日常维护，该网站目前已经在各大搜索引擎的“大数据”关键词排名中位居第一；为了解决大数据师资匮乏的问题，我面向全国院校陆续举办多期大数据师资培训班。2016 年末至今，在南京多次举办全国高校/高职/中职大数据免费培训班，基于《大数据》《大数据实验手册》以及云创大数据提供的大数据实验平台，帮助到场老师们跑通了 Hadoop、Spark 等多个大数据实验，使他们跨过了“从理论到实践，从知道到用过”的门槛。2017 年 5 月，还举办了全国千所高校大数据师资免费讲习班，盛况空前。

其中，为了解决大数据实验难的问题而开发的大数据实验平台，正在为越来越多高校的教学科研带去方便：2016 年，我带领云创大数据（www.cstor.cn，股票代码：835305）的科研人员，应用 Docker 容器技术，成功开发了 BDRack 大数据实验一体机，它打破虚拟化技术的性能瓶颈，可以为每一位参加实验的人员虚拟出 Hadoop 集群、Spark 集群、Storm 集群等，自带实验所需数据，并准备了详细的实验手册（包含 42 个大数据实验）、PPT 和实验过程视频，可以开展大数据管理、大数据挖掘等各类实验，并可进行精确营销、信用分析等多种实战演练。目前，大数据实验平台已经在郑州大学、西京学院、郑州升达经贸管理学院、镇江高等职业技术学校等多所院校成功应用，并广受校方好评。该平台也以云服务的方式在线提供（大数据实验平台，https://bd.cstor.cn），帮助师生通过自学，用一个月左右成为大数据动手的高手。

同时，为了解决缺乏权威大数据教材的问题，我所负责的南京大数据研究院，联合金陵科技学院、河南大学、云创大数据、中国地震局等多家单位，历时两年，编著出版了适合本科教学的《大数据》《大数据库》《大数据实验手册》等教材。另外，《数据挖掘》《虚拟化与容器》《大数据可视化》《深度学习》等本科教材也将于近期出版。在大数据教学中，本科院校的实践教学应更加系统性，偏向新技术的应用，且对工程实践能力要求更高。而高职、高专院校则更偏向于技术性和技能训练，理论以够用为主，学生将主要从事数据清洗和运维方面的工作。基于此，我们还联合多家高职院校专家准备了《云计算基础》《大数据基础》《数据挖掘基础》《R 语言》《数据清洗》《大数据系统运维》《大数据实践》系列教材，目前也已经陆续进入定稿出版阶段。

此外，我们也将继续在中国大数据（thebigdata.cn）和中国云计算（chinacloud.cn）等网站免费提供配套 PPT 和其他资料。同时，持续开放大数据实验平台（https://bd.cstor.cn）、免费的物联网大数据托管平台万物云（wanwuyun.com）和环境大数据免费分享平台环境云（envicloud.cn），使资源与数据随手可得，让大数据学习变得更加轻松。

在此，特别感谢我的硕士导师谢希仁教授和博士导师李三立院士。谢希仁教授所著的《计算机网络》已经更新到第 7 版，与时俱进且日臻完美，时时提醒学生要以这样的标准来写书。李三立院士是留苏博士，为我国计算机事业做出了杰出贡献，曾任国家攀登计划项目首席科学家。他的严谨治学带出了一大批杰出的学生。

本丛书是集体智慧的结晶，在此谨向付出辛勤劳动的各位作者致敬！书中难免会有不当之处，请读者不吝赐教。我的邮箱：gloud@126.com，微信公众号：刘鹏看未来（lpoutlook）。

刘鹏　教授
于南京大数据研究院

前　言

教材是体现教学内容和教学方法的知识载体，是教师授课和学生学习的重要参考资料，直接关系到教学质量和人才培养目标的实现，在教学过程中占据十分重要的地位。特别是在大数据教学中，除了理论学习外，实验尤为重要。对于大数据专业毕业生而言，拥有实际操作技能与工作经验俨然成为了其入职薪酬的加分项。以 Hadoop 开发工程师为例，Hadoop 入门月薪可达 8 千元，而具有 2~3 年工作经验的 Hadoop 人才年薪则可达到 30-50 万元。所以，大数据实验与实训直接关系到学生们的职业前景，重要性可见一斑。

然而，对于大数据实验而言，各大高校在开设课程的过程中却遇到了诸多问题。首先，大数据专业处于起步阶段，人才培养课程体系缺乏系统性，大数据教学资源匮乏，可配置和指导实验环境的专业师资不足；其次，教学过程中缺乏相应的实训项目，只有理论教育，难以培养实用型人才，存在专业学习与实际应用脱轨的情况；最后，缺乏相应的基础实验环境，无法为每一个学生都提供一套实验集群。

针对大数据实验课程建设的三大难题，我们的大数据研发团队通过长期的研究，经过反复的验证，推出了《大数据实验手册》这本教材。本教材紧扣应用型人才培养需求，本着“有用、够用、实用”的原则，在某些知识点上做了适当的扩充和提高，在突出重点、有效化解难点方面做了认真考虑和合理安排。教材打破纸上谈兵的传统模式，设计了大量的大数据实验项目，使纸质教材的实际功能辐射到学生实际操作中，引导学生对教材某些内容与观点进行探究。

本教材以实战方式进行编写，一是为了推动大数据人才培养和应用成果转化，使本书成为全国高校首选实验教材；二是为了从社会发展与高校教材发展的关系出发，寻求适应新世纪“创新人才”培养目标的新思路。同时，我们的团队开发了大数据实验平台和大数据实验一体机，可提升高校信息化管理水平和实验项目研究水平，为高校大数据课程提供基础实验环境和实验数据。

本书是集体智慧的结晶，在此谨向付出辛勤劳动的各位作者致敬！书中难免会有不当之处，请读者不吝赐教。我的邮箱：gloud@126.com，微信公众号：刘鹏看未来（lpoutlook）。

刘鹏　教授
于南京大数据研究院
2017 年 6 月 6 日

目 录

实验一　大数据实验一体机基础操作

1.1　实验目的

1．熟悉大数据实验一体机并了解如何搭建集群；
2．熟悉 Linux 基本命令；
3．掌握 vi 编辑器的使用；
4．了解 SSH 免密登录的原理以及为何需要配置 SSH 免密登录；
5．掌握如何配置 SSH 免密登录；
6．掌握 Java 基本命令；
7．熟悉集成开发软件 Eclipse 的安装和使用。

1.2　实验要求

本次实验完成后，要求学生能够：
1．使用大数据实验一体机搭建自己的集群；
2．通过 SSH 工具登录集群服务器；
3．实现每台服务器相互之间的免密登录；
4．通过 vi 编辑器编写 Java 程序；
5．通过 Java 命令编译和运行编写的 Java 程序；
6．通过 jar 命令打包编写的 Java 程序；
7．安装 Eclipse 并在其中编写 Java 程序。

1.3　实验原理

1.3.1　大数据实验一体机

随着移动互联网、云计算、物联网的快速发展，特别是智能手机端博客、社交网络、位置服务（LBS）等信息发布方式的不断涌现，数据正以前所未有的速度不断增长和累积，大数据时代已经来到。

在海量数据面前，大数据人才无疑是其中最关键环节之一。然而，不论国内外，大数据人才却相当稀缺。例如，当前我国数据人才缺口高达 150 万，而在未来 5～10 年，随着市场规模不断增加，这一缺口还将不断加大。

在创新探索大数据教学面前，高校却碰到了一系列困难，如大部分高校大数据课程体系并不完善，在实验环节，由于缺乏实验设备和大数据实训案例匮乏，实验难以开展。

针对大数据专业建设的三大难题，云创大数据为各大高校量身定制了大数据软硬件一体化的教学科研平台——大数据实验一体机。大数据实验一体机通过应用容器技术，以少量机器虚拟大量实验集群，可供大量学生同时拥有多套集群进行实验，而每个学生的实验环境不仅相互隔离，方便高效地完成实验，而且实验彼此不干扰，即使某个实验环境被破坏，对其他人也没有影响，一键重启就可以拥有一套新集群，大幅度节省了硬件和人员管理的投入成本。

此外，作为一个可供大量学生完成大数据与云计算实验的集成环境，该平台同步提供了配套的培训服务，对于教学组件的安装、配置，教材、实验手册等具体应用提供一站式服务，有助于更好地满足高校课程设计、课程上机实验、实习实训、科研训练等多方面需求，并在一定程度上缓解大数据师资不足的问题。对于各大高校而言，即使没有任何大数据实验基础，该平台也能助其轻松开展大数据与云计算的教学、实验与科研。

具体而言，大数据实验一体机从以下四个方面解决了高校大数据的教学科研难题。

（1）完整的大数据课程体系及配套资源，一步解决入门难的问题

在《实战 Hadoop2.0——从云计算到大数据》和实验手册的指导之下，大数据实验一体机解决方案涵盖大数据算法、接口、工具、平台等多方面内容，从大数据监测与收集、大数据存储与处理、大数据分析与挖掘直至大数据创新，帮助高校构建完整的大数据课程体系。

综合 36 个大数据实验的实验手册及配套高清视频课程，涵盖原理验证、综合应用、自主设计及创新的多层次实验内容。每个实验呈现详细的实验目的、实验内容、实验原理和实验流程指导。配套相应的实验数据和高清视频课程，参照手册即可轻松完成每个实验。中国大数据、中国云计算、中国存储等国内大数据和云计算专业领域排名第一的网站将会提供全线支持，一网打尽各类优质资源。

（2）安全可靠的实验环境，大幅度提升大数据技能

基于 Docker 容器技术，大数据实验一体机可快速创建随时运行的实验环境。使用几台机器即可虚拟出大量实验集群，方便上百学生同时使用。采用 Kubernetes+ZooKeeper 架构管理集群，实验集群完全隔离。实验环境互不干扰，如果实验环境被破坏，一键重启即可建立新集群。内置数据挖掘等教学实验数据，可导入高校各学科数据进行教学、科研，校外培训机构同样适用。

（3）热门实战项目贯穿始终，进一步提高教学效果与就业率

大数据实验一体机解决方案采用理论与实践相结合的人才培养模式，帮助教师提高教学水平，促使学生完善大数据知识体系。基于真实的企业基地实训经验，提供丰富的项目实训案例。结合高校各专业实际情况进行行业数据研究，培养实用型人才的专业项目能力。

（4）更多潜在效益，同步增强高校的硬实力和影响力

大数据上升为国家战略，发改委明确组建 13 个国家级大数据实验室，大数据实验一体机有助于高校大数据实验室建设以及高层次大数据人才的深度培育。大数据实验一体机解决方案在理论与实践双管齐下，帮助提升了高校信息化管理水平和实验项目研究水平。大数据产业迎来发展黄金期，大数据实验一体机可提高大数据专业就业率，进一步增强高校的硬实力和影响力。

在 2016 年暑期全国高校大数据培训中，云创大数据利用大数据实验一体机搭建了 Docker 容器云，为每个学员分配 5 套虚拟服务器集群，提供了简洁易用的上机操作环境，得到了学员的一致好评。在理论讲解的基础上，讲师通过这一实践平台，为学员提供精确到每一步的操作指导，真正做到了学思结合、知行统一，所有学员的大数据应用能力均得以提升，并获得了相应的大数据能力等级证书。

大数据实验一体机基本操作主要包括账号管理、集群管理、集群登录和辅助功能四大部分，其中账号管理完成新建和销毁用户账号，集群管理完成新建和销毁集群，集群登录指通过 SSH 登录到集群各机器，辅助功能模板提供了部分软件下载等实用小功能。

1. 界面管理

输入本校大数据实验一体机网址后，请输入相应账号与密码，点击登录即可。如图 1-1 所示。

图 1-1　登录界面

2. 账号管理

系统管理员和教师角色登录后，可以看到用户账号管理界面。

系统管理员用户可以在该界面中查看或修改所有的教师和学生用户信息，并可以注册、销毁教师或学生用户账户；

教师用户可以在该界面中查看或修改自己建立的所有学生用户信息，并可以注册或销毁自己的学生用户账户，如图 1-2 所示。

图 1-2　账号管理

3. 集群管理

此处的集群管理包含创建集群和销毁集群，由于云创大数据实验一体机采用 Docker 技术，因此能够在几乎不占用系统资源情况下，实现大量机器快速创建与销毁，不必担心资源消耗高、启动销毁慢、管理维护难等问题。

（1）创建集群

当需要新建集群时，直接点击集群管理界面的创建集群即可，后台会快速为用户新建五台预安装 CentOS 7 操作系统的机器，并配置好各自的主机名和 IP 地址等。如图 1-3 所示。

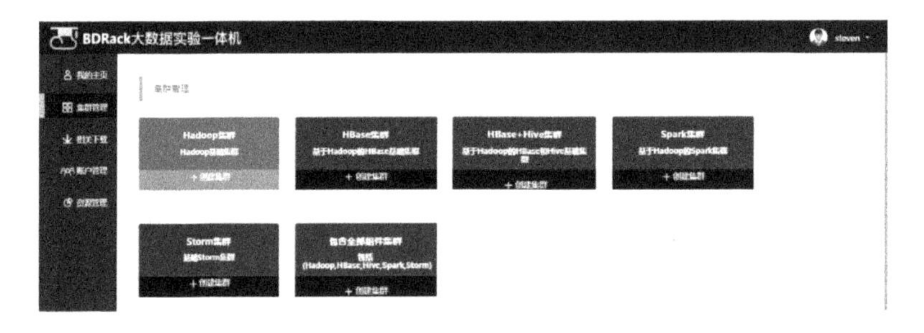

图 1-3　创建集群

（2）销毁集群

若实验过程中，由于命令敲错等各种原因导致集群无法使用，可在“我的主页”中随时销毁失效的集群，之后再重新建立新的集群。集群主页如图 1-4 所示。

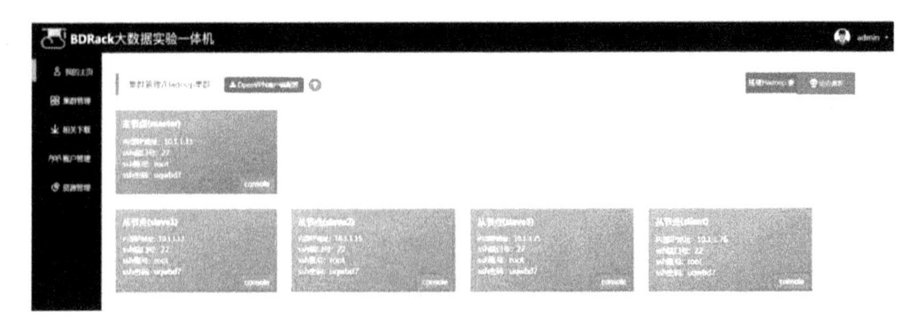

图 1-4　集群主页

4. 相关下载

大数据实验一体机的相关下载界面提供了实验所需的软件及插件的下载，为避免软件版本不同导致实验环境配置错误，请尽量下载和使用此处指定的软件版本与插件。如图 1-5 所示。

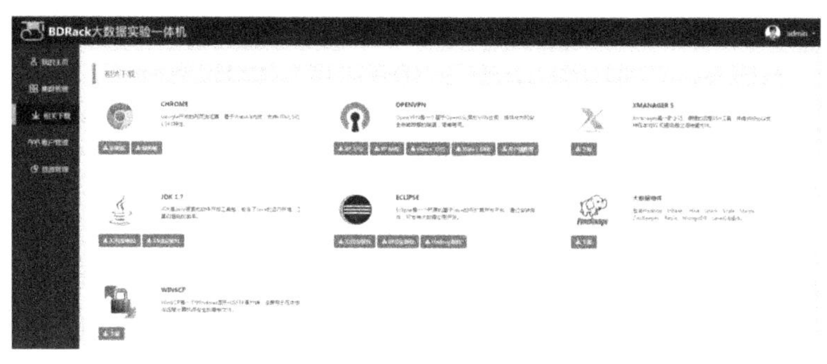

图 1-5　相关下载

1.3.2　Linux 基本命令

云创大数据实验平台搭建的集群服务器均为预装 Linux 操作系统的服务器。

Linux 是一套免费使用和自由传播的类 UNIX 操作系统，是一个基于 POSIX 和 UNIX 的多用户、多任务、支持多线程和多 CPU 的操作系统。它能运行主要的 UNIX 工具软件、应用程序和网络协议。它支持 32 位和 64 位硬件。Linux 继承了 UNIX 以网络为核心的设计思想，是一个性能稳定的多用户网络操作系统。

Linux 操作系统于 1991 年 10 月 5 日成立。Linux 存在着许多不同的 Linux 版本，但它们都使用了 Linux 内核。Linux 可安装在各种计算机硬件设备中，例如手机、平板电脑、路由器、视频游戏控制台、台式计算机、大型机和超级计算机。

严格来讲，Linux 这个词本身只表示 Linux 内核，但实际上人们已经习惯了用 Linux 来形容整个基于 Linux 内核，并且使用 GNU 工程各种工具和数据库的操作系统。

木小节将介绍实验中涉及的 Linux 操作系统命令。

（1）查看当前目录

pwd 命令用于显示当前目录：

```
[root@master ~]# pwd
/root
```

（2）目录切换

cd 命令用来切换目录：

```
[root@master ~]# cd  /usr/cstor
[root@master cstor]# pwd
/usr/cstor
[root@master cstor]#
```

（3）文件罗列

ls 命令用于查看文件与目录：

```
[root@master cstor]# ls
```

（4）文件或目录复制

cp 命令用于复制文件，若复制的对象为目录，则需要使用-r 参数：

```
[root@master cstor]#   cp   -r hadoop   /root/hadoop
```

（5）文件或目录移动或重命名

mv 命令用于移动文件，在实际使用中，也常用于重命名文件或目录：

```
[root@master ~]#   mv   hadoop   hadoop2                          #当前位于/root，不是
/usr/cstor
```

（6）文件或目录删除

rm 命令用于删除文件，若删除的对象为目录，则需要使用-r 参数：

```
[root@master ~]#   rm   -rf   hadoop2                              #当前位于/root，不是
/usr/cstor
```

（7）进程查看

ps 命令用于查看系统的所有进程：

```
[root@master ~]# ps                                   # 查看当前进程
```

（8）文件压缩与解压

tar 命令用于文件压缩与解压，参数中的 c 表示压缩，x 表示解压缩：

```
[root@master ~]# tar -zcvf   /root/hadoop.tar.gz   /usr/cstor/hadoop
[root@master ~]# tar -zxvf   /root/hadoop.tar.gz
```

（9）查看文件内容

cat 命令用于查看文件内容：

```
[root@master ~]# cat   /usr/cstor/hadoop/etc/hadoop/core-site.xml
```

（10）查看服务器 IP 配置

ip addr 命令用于查看服务器 IP 配置：

```
[root@master ~]# ip addr
1: lo: <LOOPBACK,UP,LOWER_UP> mtu 65536 qdisc noqueue state UNKNOWN
    link/loopback 00:00:00:00:00:00 brd 00:00:00:00:00:00
    inet 127.0.0.1/8 scope host lo
        valid_lft forever preferred_lft forever
    inet6 ::1/128 scope host
        valid_lft forever preferred_lft forever
125: eth0@if126: <BROADCAST,MULTICAST,UP,LOWER_UP> mtu 1500 qdisc noqueue state UP
    link/ether 02:42:ac:11:00:0c brd ff:ff:ff:ff:ff:ff link-netnsid 0
    inet 172.17.0.12/16 scope global eth0
        valid_lft forever preferred_lft forever
    inet6 fe80::42:acff:fe11:c/64 scope link
        valid_lft forever preferred_lft forever
[root@master ~]#
```

1.3.3 vi 编辑器

vi 编辑器通常被简称为 vi，而 vi 又是 visual editor 的简称。它在 Linux 上的地位就

像 Edit 程序在 DOS 上一样。它可以执行输出、删除、查找、替换、块操作等众多文本操作，而且用户可以根据自己的需要对其进行定制，这是其他编辑程序所没有的。

vi 编辑器并不是一个排版程序，它不像 Word 或 WPS 那样可以对字体、格式、段落等其他属性进行编排，它只是一个文本编辑程序，没有菜单，只有命令，且命令繁多。vi 有三种基本工作模式：命令行模式、文本输入模式和末行模式。

vim 是 vi 的加强版，比 vi 更容易使用。vi 的命令几乎全部都可以在 vim 上使用。

vi 编辑器是 Linux 和 UNIX 上最基本的文本编辑器，工作在字符模式下。由于不需要图形界面，vi 是效率很高的文本编辑器。尽管在 Linux 上也有很多图形界面的编辑器可用，但 vi 在系统和服务器管理中的功能是那些图形编辑器所无法比拟的。

vi 或 vim 是实验中用到最多的文件编辑命令，命令行嵌入"vi/vim 文件名"后，默认进入"命令模式"，不可编辑文档，需键盘点击"i"键，方可编辑文档，编辑结束后，需按"ESC"键，先退回命令模式，再按"："进入末行模式，接着嵌入"wq"方可保存退出。图 1-6 为 vi/vim 三种模式转换，图 1-7 为 vi/vim 操作实例。

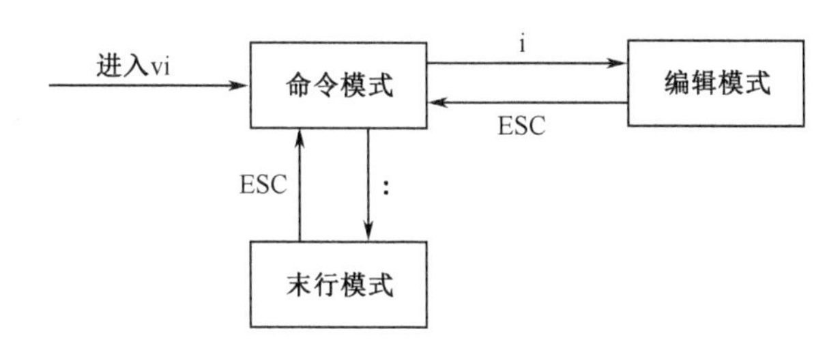

图 1-6　vi/vim 三种模式转换

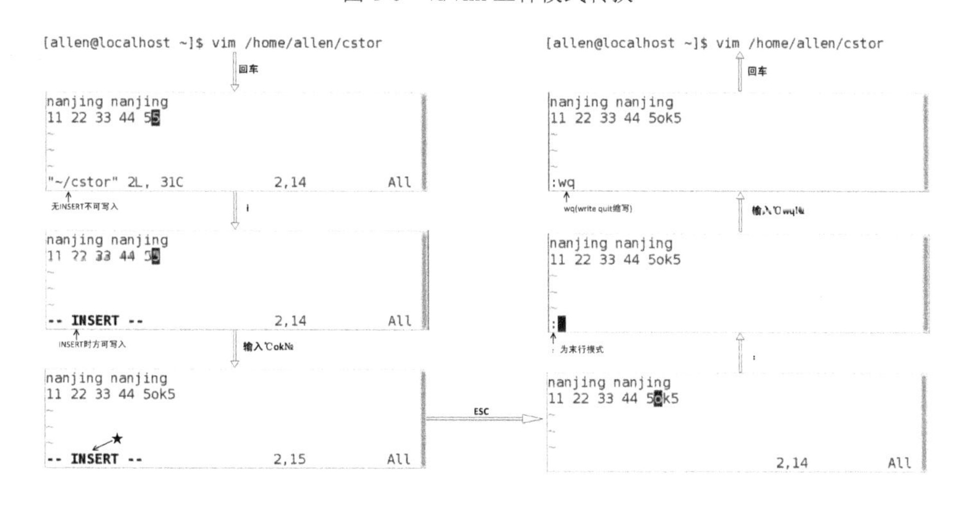

图 1-7　vi/vim 操作实例

1.3.4　SSH 免密认证

实验中，我们需要从实验室机器登录到集群中的 Linux 服务器上，而绝大多数 Linux 服务器采用的是 SSH（Secure Shell）登录方式，因此，我们需要在实验室机器上安装一个 SSH 登录工具。常用的 SSH 工具包括 XShell、Secure CRT、putty 等，大数据实验一体机的相关下载界面中提供了 XShell 工具的下载。

Hadoop 的基础是分布式文件系统 HDFS，HDFS 集群有两类节点以管理者-工作者的模式运行，即一个 namenode（管理者）和多个 datanode（工作者）。在 Hadoop 启动以后，namenode 通过 SSH 来启动和停止各个节点上的各种守护进程，这就需要在这些节点之间执行指令时采用无须输入密码的认证方式，因此，我们需要将 SSH 配置成使用无须输入 root 密码的密钥文件认证方式，如图 1-8 所示。

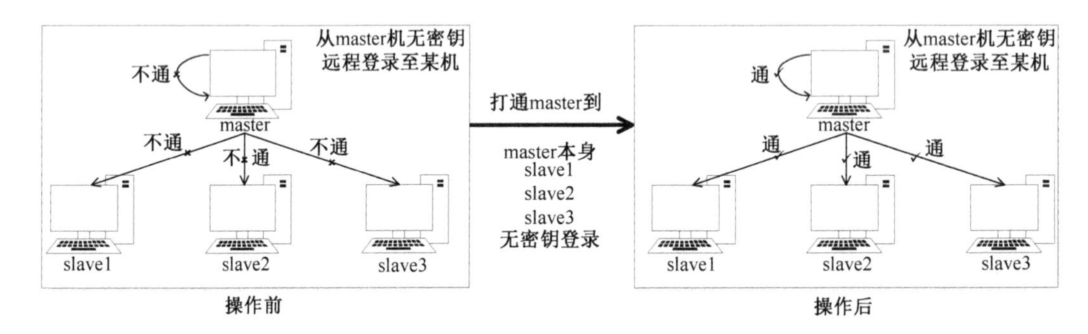

图 1-8　实验集群 master 服务器 SSH 免密登录

1.3.5　Java 基本命令

在安装 Java 环境后，可以使用 Java 命令来编译、运行或者打包 Java 程序。

（1）查看 Java 版本

```
[root@client ~]# java -version
java version "1.7.0_79"
Java(TM) SE Runtime Environment (build 1.7.0_79-b15)
Java HotSpot(TM) 64-Bit Server VM (build 24.79-b02, mixed mode)
```

（2）编译 Java 程序

```
[root@client ~]# javac Helloworld.java
```

（3）运行 Java 程序

```
[root@client ~]# java Helloworld
Hello World!
```

（4）打包 Java 程序

```
[root@client ~]# jar -cvf Helloworld.jar Helloworld.class
added manifest
adding: Helloworld.class(in = 426) (out= 289)(deflated 32%)
```

由于打包时并没有指定 manifest 文件，因此该 jar 包无法直接运行：

```
[root@client ~]# java -jar Helloworld.jar
no main manifest attribute, in Helloworld.jar
```

（5）打包携带 manifest 文件的 Java 程序

manifest 文件用于描述整个 Java 项目，最常用的功能是指定项目的入口类：

```
[root@client ~]# cat manifest.mf
Main-Class: Helloworld
```

打包时，加入-m 参数，并指定 manifest 文件名：

```
[root@client ~]# jar -cvfm Helloworld.jar manifest.mf Helloworld.class
added manifest
adding: Helloworld.class(in = 426) (out= 289)(deflated 32%)
```

之后，即可使用 java 命令直接运行该 jar 包：

```
[root@client ~]# java -jar Helloworld.jar
Hello World!
```

1.3.6　Eclipse 集成开发环境

Eclipse 是一个开放源代码的、基于 Java 的可扩展开发平台。就其本身而言，它只是一个框架和一组服务，用于通过插件组件构建开发环境。幸运的是，Eclipse 附带了一个标准的插件集，包括 Java 开发工具（Java Development Kit，JDK）。

Eclipse 是著名的跨平台的自由集成开发环境（IDE）。最初主要用来 Java 语言开发，通过安装不同的插件 Eclipse 可以支持不同的计算机语言，比如 C++和 Python 等开发工具。Eclipse 的本身只是一个框架平台，但是众多插件的支持使得 Eclipse 拥有其他功能相对固定的 IDE 软件很难具有的灵活性。许多软件开发商以 Eclipse 为框架开发自己的 IDE。

使用 Eclipse 可以帮助程序开发人员自动补全语义、方法名、方法参数、语句块等，并且能够实时检查程序语法，提供错误和警告说明等，极大地提高了开发效率。

然而，使用 Eclipse 会占用较大的系统内存，因此，对于配置不高（32 位操作系统或内存不足 4G）的实验机器，不推荐安装 Eclipse。

1.4　实验步骤

1.4.1　搭建集群服务器

使用自己的账号密码登录大数据实验一体机（默认密码为 123456，登录后会自动跳转至密码修改界面，建议修改为自己的密码），进入集群管理界面，如图 1-9 所示。

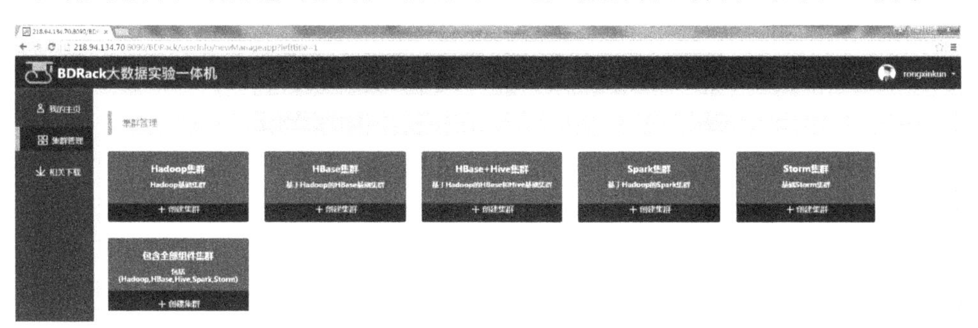

图 1-9　登录大数据实验一体机

选择第一个 Hadoop 集群，点击创建集群，等待集群建立完成，如图 1-10 所示。

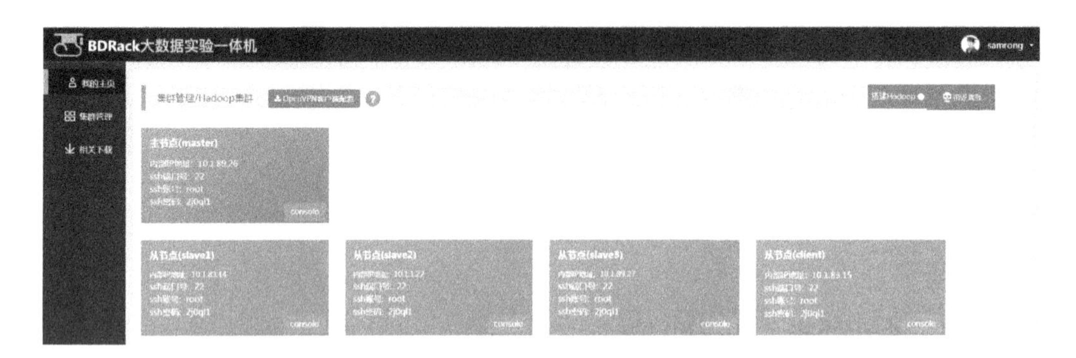

图 1-10　搭建 Hadoop 集群完成

1.4.2　使用 SSH 工具登录每台服务器

在搭建好的 Hadoop 集群中，已经给出了所有五台服务器的内部 IP 地址、SSH 端口号、SSH 登录名以及 SSH 登录密码。

要想登录这些服务器，我们需要先下载 OpenVPN 客户端软件。

在相关下载中，根据 PC 的操作系统版本下载对应版本的 OpenVPN 客户端安装包，并下载客户端配置文件，如图 1-11 所示。

图 1-11　下载 OpenVPN

安装完成后，将下载的客户端配置压缩包解压，将其中的 client.ovpn 放于 OpenVPN 安装目录的 config 文件夹下。

以管理员身份运行 OpenVPN GUI，任务栏将出现 OpenVPN GUI 图标，右键单击任务栏内 OpenVPN GUI 图标，点击“Connect”，如图 1-12 所示。

图 1-12　登录集群服务器（一）

当提示连接成功后，即可使用 SSH 工具登录大数据试验一体机分配的内网 IP 连接你的集群服务器。如图 1-13 所示。

```
master | slave1 | slave2 | slave3 | client ×
[root@client ~]#
```

图 1-13　登录集群服务器（二）

1.4.3　添加域名映射

系统搭建好的集群服务器已经完成修改主机名、关闭防火墙、安装 JDK、同步时钟四步操作，为了可以安装大数据组件，还需为所有机器添加域名映射，如图 1-14 所示。

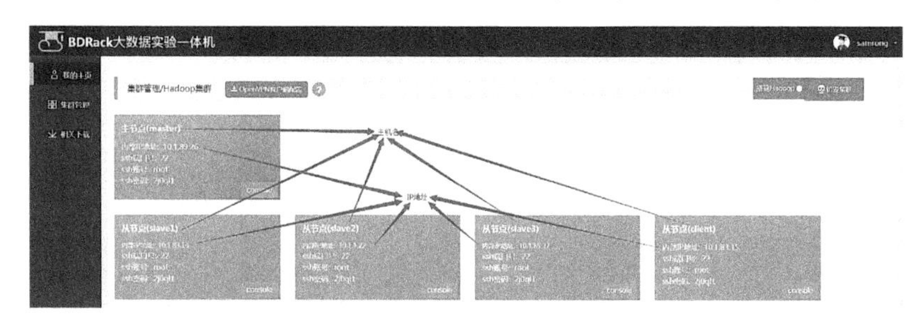

图 1-14　汇总各机机器名及其对应 IP

使用 SSH 工具登录到 master 服务器，使用 vi 命令编辑/etc/hosts 文件：

```
[root@master ~]#   vi   /etc/hosts                    #root 权限,编辑 master 的域名映射文件
```

在文件的末尾追加写入如下五行（具体的 IP 地址请替换为实际集群服务器内部 ip）：

```
172.17.0.7      master
172.17.0.10     slave1
172.17.0.33     slave2
172.17.0.8      slave3
172.17.0.34     client
```

保存退出后，master 服务器的域名映射即添加完成，使用 cat 命令查看/etc/hosts 文件。如图 1-15 所示。

```
[root@master ~]# cat  /etc/hosts
127.0.0.1       localhost
::1     localhost ip6-localhost ip6-loopback      ┐
fe00::0 ip6-localnet                              ├ 请勿删除
ff00::0 ip6-mcastprefix                           │
ff02::1 ip6-allnodes                              ┘
ff02::2 ip6-allrouters
172.17.0.7      master  ←┐
172.17.0.7    master    ←┴ 有重复项,无影响
172.17.0.10   slave1    ┐
172.17.0.33   slave2    ├ 追加的内容
172.17.0.8    slave3    │
172.17.0.34   client    ┘
[root@master ~]#
```

图 1-15　配置域名映射文件

依次登录 slave1~slave3 和 client 服务器，重复该操作。

1.4.4 配置 SSH 免密登录

1. 生成 master 服务器密钥

执行命令 ssh-keygen，生成 master 服务器密钥，如图 1-16 所示。

```
[root@master ~]#  ssh-keygen                                    #root 用户,master 机,生成公私钥
```

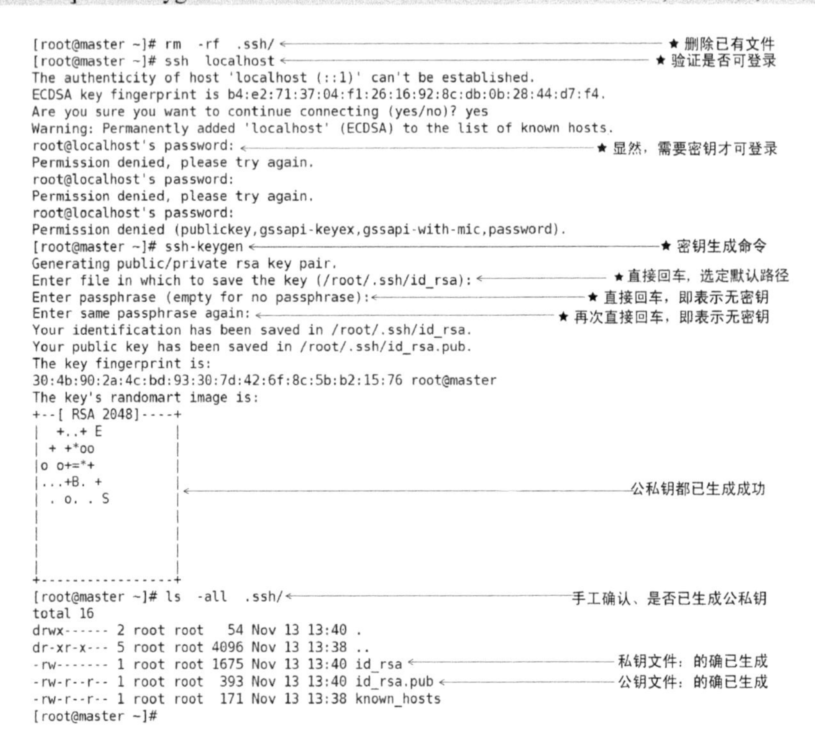

图 1-16 master 密钥生成

如图 1-16 所示，在 master 上执行“ssh-keygen”命令生成公私钥。第一个提示是询问将公私钥文件存放在哪，直接回车，选择默认位置。第二个提示是请求用户输入密钥，既然操作的目的就是实现 SSH 无密钥登录，故此处必须使用空密钥，所谓的空密钥指的是直接回车，不是空格，更不是其他字符。此处请读者务必直接回车，使用空密钥。第三个提示是要求用户确认刚才输入的密钥，既然刚才是空密钥（直接回车即空），那现在也应为空，直接回车即可。

最后，可通过命令“ls -all /root/.ssh”查看到，SSH 密钥文件夹.ssh 目录下的确生成了两个文件 id_rsa 和 id_rsa_pub，这两个文件都有用，其中公钥用于加密，私钥用于解密。中间的 rsa 表示算法为 RSA 算法。

2. 复制 master 服务器公钥至本机

执行命令 ssh-copy-id master，将 master 服务器公钥复制至 master 服务器本身。如图 1-17 所示。

```
[root@master ~]# ssh-copy-id master
The authenticity of host 'master (172.17.0.12)' can't be established.
ECDSA key fingerprint is b4:e2:71:37:04:f1:26:16:92:8c:db:0b:28:44:d7:f4.
Are you sure you want to continue connecting (yes/no)? yes
/usr/bin/ssh-copy-id: INFO: attempting to log in with the new key(s), to filter out any that are already installed
/usr/bin/ssh-copy-id: INFO: 1 key(s) remain to be installed -- if you are prompted now it is to install the new keys
root@master's password:

Number of key(s) added: 1

Now try logging into the machine, with:   "ssh 'master'"
and check to make sure that only the key(s) you wanted were added.

[root@master ~]#
```

图 1-17　复制 master 公钥至 master

第一次连接 master 时，需要输入 yes 来确认建立授权的主机名访问，并需要输入 root 用户密码来完成公钥文件传输。

3. 验证 master 服务器 SSH 免密登录 master 本身

公钥复制完成后，可以在 master 服务器上直接执行命令 ssh master，查看是否可以免密登录 master 服务器：

```
[root@master ~]# ssh master          #root 用户,登录本机网络地址
[root@master ~]# exit                #退出本次登录
logout
Connection to master closed.
[root@master ~]#
```

4. 复制 master 服务器公钥至其余服务器

执行命令 ssh-copy-id slave1，将 master 服务器公钥复制至 slave1 服务器，如图 1-18 所示。

```
master × | slave1 | slave2 | slave3 | client
[root@master ~]# ssh-copy-id slave1
The authenticity of host 'slave1 (172.17.0.14)' can't be established.
ECDSA key fingerprint is b4:e2:71:37:04:f1:26:16:92:8c:db:0b:28:44:d7:f4.
Are you sure you want to continue connecting (yes/no)? yes
/usr/bin/ssh-copy-id: INFO: attempting to log in with the new key(s), to filter out any that are already installed
/usr/bin/ssh-copy-id: INFO: 1 key(s) remain to be installed -- if you are prompted now it is to install the new keys
root@slave1's password:

Number of key(s) added: 1

Now try logging into the machine, with:   "ssh 'slave1'"
and check to make sure that only the key(s) you wanted were added.

[root@master ~]#
```

图 1-18　复制 master 公钥至 slave1

第一次连接 slave1 时，需要输入 yes 来确认建立授权的主机名访问，并需要输入 root 用户密码来完成公钥文件传输。

依照同样的方式将公钥复制至 slave2、slave3 和 client 服务器。

5. 验证 master 服务器 SSH 免密登录其他服务器

公钥复制完成后，可以在 master 服务器上直接执行命令 ssh master，查看是否可以免密登录 slave1~slave3 和 client 服务器：

```
[root@master ~]# ssh  localhost          #root 用户,登录本机环回地址
[root@master ~]# ssh  master             #root 用户,登录本机网络地址
[root@master ~]# ssh  slave1             #root 用户,从 master 远程登录 slave1
[root@master ~]# ssh  slave2             #root 用户,从 master 远程登录 slave2
```

```
[root@master ~]# ssh    slave3                                    #root 用户,从 master 远程登录 slave3
```

6. 其余服务器配置 SSH 免密登录

其余服务器按照同样的方式配置 SSH 免密登录，完成后验证是否可以互相之间实现 SSH 免密登录。

1.4.5 在 client 服务器开发 Java Helloworld 程序

使用 SSH 工具登录 client 服务器，使用 vi 编辑器编写 Helloworld.java：

```
public class Helloworld {
    public static void main(String args[]) {
        System.out.println("Hello World!");
    }
}
```

使用 Javac 命令编译该程序，生成 Helloworld.class 文件：

```
[root@client ~]# javac Helloworld.java
[root@client ~]# ls
anaconda-ks.cfg   data   envSource   Helloworld.class   Helloworld.java
```

使用 Java 命令执行该程序，输出 Hello World!

```
[root@client ~]# java Helloworld
Hello World!
```

1.4.6 使用 Eclipse 开发 Java Helloworld 程序

根据实验室机器的环境，下载并安装对应版本的 JDK 和 Eclipse 软件（若已安装则跳过该步骤）。

安装完成后，双击 Eclipse 图标，打开该软件，其界面如图 1-19 所示。

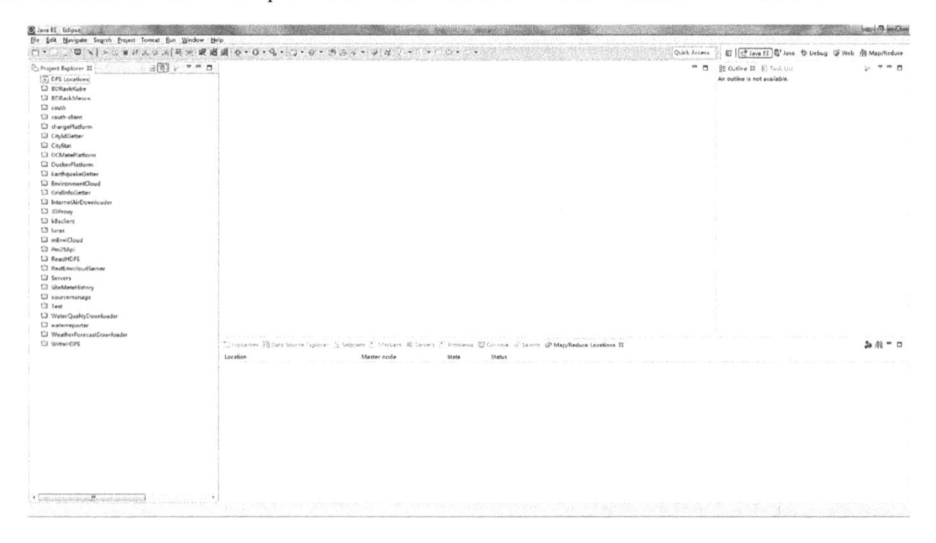

图 1-19　Eclipse 界面

依次点击 File→New→Java Project 或 File→New→Other→Java Project，项目名为

Demo，如图 1-20 所示。

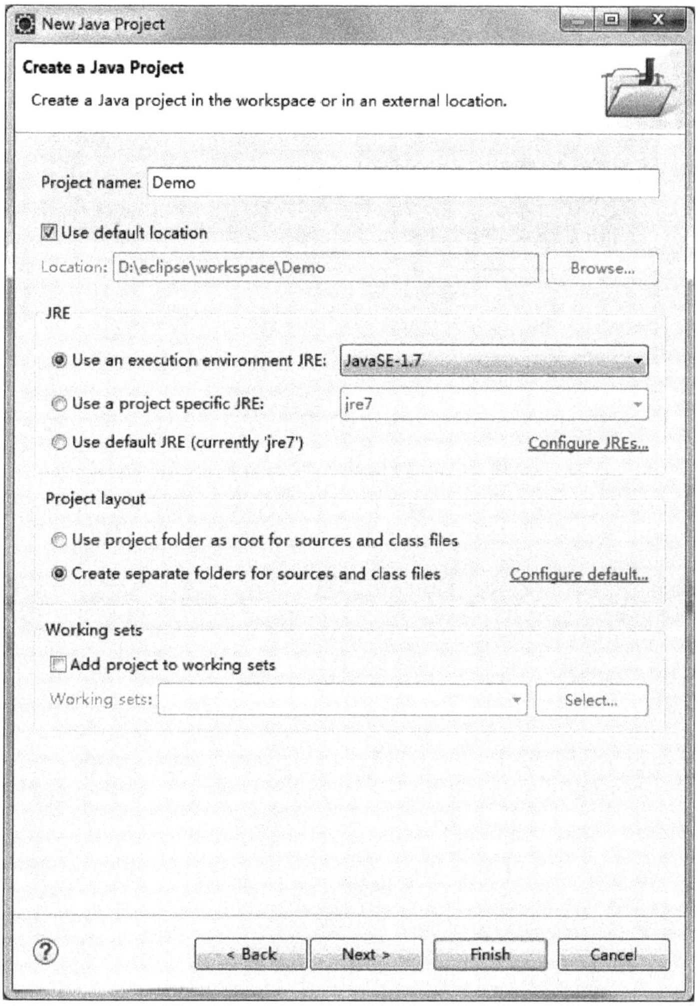

图 1-20　创建 Java 项目

点击 Finish，新建项目完成。在左侧导航栏中选中该项目，右键点击，选择 New→Class，新建 Helloworld 类，并在该文件入如下内容：

```
public class Helloworld {
    public static void main(String args[]) {
        System.out.println("Hello World!");
    }
}
```

完成后，点击上方的 Run 按钮，即可执行该程序，如图 1-21 所示。

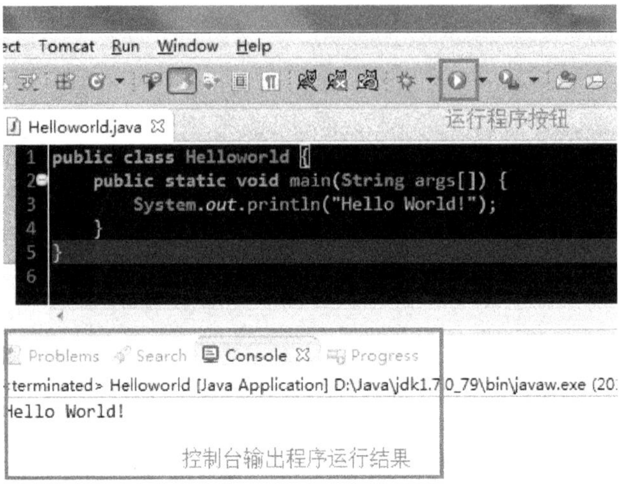

图 1-21 Eclipse 执行 Java 程序

实验二　HDFS 实验：部署 HDFS

2.1　实验目的

1．理解 HDFS 存在的原因；
2．理解 HDFS 体系架构；
3．理解 master/slave 架构；
4．理解为何配置文件里只需指定主服务，无须指定从服务；
5．理解为何需要客户端节点；
6．学会逐一启动 HDFS 和统一启动 HDFS；
7．学会在 HDFS 中上传文件。

2.2　实验要求

要求实验结束时，已构建出以下 HDFS 集群：
1．master 上部署主服务 NameNode；
2．slave1、slave2、slave3 上部署从服务 DataNode；
3．client 上部署 HDFS 客户端。
待集群搭建好后，还需在 client 上进行下述操作：
1．在 HDFS 里新建目录；
2．将 client 上某文件上传至 HDFS 里刚才新建的目录。

2.3　实验原理

2.3.1　分布式文件系统

分布式文件系统（Distributed File System）是指文件系统管理的物理存储资源不一定直接连接在本地节点上，而是通过计算机网络与节点相连。该系统架构于网络之上，势必会引入网络编程的复杂性，因此分布式文件系统比普通磁盘文件系统更为复杂。

2.3.2　HDFS

HDFS（Hadoop Distributed File System）为大数据平台其他所有组件提供了最基本的存储功能。它具有高容错、高可靠、可扩展、高吞吐率等特征，为大数据存储和处理提供了强大的底层存储架构。

HDFS 是一个主/从（master/slave）体系结构，从最终用户的角度来看，它就像传统的文件系统，可通过目录路径对文件执行 CRUD 操作。由于其分布式存储的性质，HDFS 集群拥有一个 NameNode 和一些 DataNodes，NameNode 管理文件系统的元数据，DataNode 存储实际的数据。

HDFS 开放文件系统的命名空间以便用户以文件形式存储数据，秉承“一次写入、多次读取”的原则。客户端通过 NameNode 和 DataNodes 的交互访问文件系统，联系 NameNode 以获取文件的元数据，而真正的文件 I/O 操作是直接和 DataNode 进行交互的。

2.3.3 HDFS 基本命令

HDFS 基本命令格式如下：

```
hadoop fs -cmd args
```

其中，cmd 为具体的操作，args 为参数。

部分 HDFS 命令示例如下：

```
hadoop fs -mkdir /user/trunk                #建立目录/user/trunk
hadoop fs -ls /user                         #查看/user 目录下的目录和文件
hadoop fs -lsr /user                        #递归查看/user 目录下的目录和文件
hadoop fs -put test.txt /user/trunk         #上传 test.txt 文件至/user/trunk
hadoop fs -get /user/trunk/test.txt         #获取/user/trunk/test.txt 文件
hadoop fs -cat /user/trunk/test.txt         #查看/user/trunk/test.txt 文件内容
hadoop fs -tail /user/trunk/test.txt        #查看/user/trunk/test.txt 文件的最后 1000 行
hadoop fs -rm /user/trunk/test.txt          #删除/user/trunk/test.txt 文件
hadoop fs -help ls                          #查看 ls 命令的帮助文档
```

2.3.4 HDFS 适用场景

HDFS 提供高吞吐量应用程序数据访问功能，适合带有大型数据集的应用程序，以下是一些常用的应用场景。

数据密集型并行计算：数据量极大，但是计算相对简单的并行处理，如大规模 Web 信息搜索；

计算密集型并行计算：数据量相对不是很大，但是计算较为复杂的并行处理，如 3D 建模与渲染、气象预报和科学计算；

数据密集与计算密集混合型的并行计算：如 3D 电影的渲染。

HDFS 在使用过程中有以下限制：

（1）HDFS 不适合大量小文件的存储，由于 NameNode 将文件系统的元数据存放在内存中，因此存储的文件数目受限于 NameNode 的内存大小；

（2）HDFS 适用于高吞吐量，而不适合低时间延迟的访问；

（3）流式读取的方式，不适合多用户写入一个文件（一个文件同时只能被一个客户端写），以及任意位置写入（不支持随机写）；

（4）HDFS 更加适合写入一次，读取多次的应用场景。

2.4 实验步骤

部署 HDFS 主要步骤如下：

（1）配置 Hadoop 的安装环境；

（2）配置 Hadoop 的配置文件；

（3）启动 HDFS 服务；

（4）验证 HDFS 服务可用。

2.4.1 在 master 服务器上确定存在 Hadoop 安装目录

```
[root@master ~]# ls   /usr/cstor/hadoop
```

显示结果如图 2-1 所示。

```
client | master × | slave1 | slave2 | slave3
[root@master ~]# ll /usr/cstor/hadoop/
total 40
drwxr-xr-x 2 10021 10021  4096 Jun 29  2015 bin
drwxr-xr-x 3 root  root     16 Dec  5 12:21 cloud
drwxr-xr-x 3 10021 10021    19 Jun 29  2015 etc
drwxr-xr-x 2 10021 10021   101 Jun 29  2015 include
drwxr-xr-x 3 10021 10021    19 Jun 29  2015 lib
drwxr-xr-x 2 10021 10021  4096 Jun 29  2015 libexec
-rw-r--r-- 1 10021 10021 15429 Jun 29  2015 LICENSE.txt
drwxr-xr-x 2 root  root   4096 Dec  5 12:22 logs
-rw-r--r-- 1 10021 10021   101 Jun 29  2015 NOTICE.txt
-rw-r--r-- 1 10021 10021  1366 Jun 29  2015 README.txt
drwxr-xr-x 2 10021 10021  4096 Jun 29  2015 sbin
drwxr-xr-x 4 10021 10021    29 Jun 29  2015 share
[root@master ~]#
```

图 2-1　确认 Hadoop 安装目录显示结果

2.4.2 确认集群服务器之间可 SSH 免密登录

使用 SSH 工具登录到每一台服务器，执行命令 ssh 主机名，确认每台集群服务器均可 SSH 免密登录。若无法 SSH 免密登录，请参照实验一的 1.4.4 节进行配置。

2.4.3 修改 HDFS 配置文件

（1）设置 JDK 安装目录

编辑文件“/usr/cstor/hadoop/etc/hadoop/hadoop-env.sh”，找到如下一行：

```
export JAVA_HOME=${JAVA_HOME}
```

将这行内容修改为：

```
export JAVA_HOME=/usr/local/jdk1.7.0_79/
```

这里的“/usr/local/jdk1.7.0_79/”就是 JDK 安装位置，如果不同，请根据实际情况更改。

（2）指定 HDFS 主节点

编辑文件“/usr/cstor/hadoop/etc/hadoop/core-site.xml”，将如下内容嵌入此文件里最后两行的<configuration></configuration>标签之间：

```
<property><name>hadoop.tmp.dir</name><value>/usr/cstor/hadoop/cloud</value></property>
<property><name>fs.defaultFS</name><value>hdfs://master:8020</value></property>
```

（3）复制集群配置至其他服务器

在 master 机上执行下列命令，将配置好的 hadoop 复制至 slaveX、client。

```
[root@master ~]# cat   ~/data/2/machines
slave1
salve2
slave3
client
[root@master ~]# for  x  in  'cat  ~/data/2/machines' ; do  echo  $x ; scp  -r  /usr/cstor/hadoop/etc
$x:/usr/cstor/hadoop  ; done;
```

2.4.4 启动 HDFS

在 master 服务器上格式化主节点：

```
[root@master ~]# hdfs  namenode  -format
```

配置 slaves 文件，将 localhost 修改为 slave1~slave3：

```
[root@master ~]# vi /usr/cstor/hadoop/etc/hadoop/slaves
slave1
slave2
slave3
```

统一启动 HDFS：

```
[root@master ~]#cd /usr/cstor/hadoop
[root@master hadoop]# sbin/start-dfs.sh
```

2.4.5 通过查看进程的方式验证 HDFS 启动成功

分别在 master、slave1~slave3 四台机器上执行如下命令，查看 HDFS 服务是否已启动。

```
[root@master sbin]#   jps                                    #jps 查看 java 进程
```

若启动成功，在 master 上会看到类似的如下信息：

```
6208 NameNode
6862 Jps
6462 SecondaryNameNode
```

而在 slave1、slave2、slave3 上会看到类似的如下信息：

```
6208 DataNode
6862 Jps
```

2.4.6 使用 client 上传文件

从 client 服务器向 HDFS 上传文件。

```
[root@client ~]# hadoop fs -put   ~/data/2/machines /
```

执行命令：hadoop fs -ls /，查看文件是否上传成功，如图 2-2 所示。

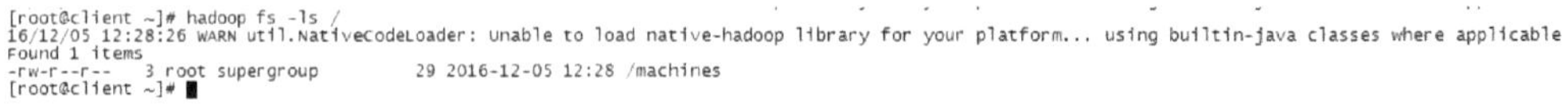

```
[root@client ~]# hadoop fs -ls /
16/12/05 12:28:26 WARN util.NativeCodeLoader: Unable to load native-hadoop library for your platform... using builtin-java classes where applicable
Found 1 items
-rw-r--r--   3 root supergroup         29 2016-12-05 12:28 /machines
[root@client ~]# 
```

图 2-2　查看文件是否上传成功

实验三　HDFS 实验：读写 HDFS 文件

3.1　实验目的

1．会在 Linux 环境下编写读写 HDFS 文件的代码；
2．会使用 jar 命令打包代码；
3．会在 client 服务器上运行 HDFS 读写程序；
4．会在 Windows 上安装 Eclipse Hadoop 插件；
5．会在 Eclipse 环境编写读写 HDFS 文件的代码；
6．会使用 Eclipse 打包代码；
7．会使用 Xftp 工具将实验电脑上的文件上传至 client 服务器。

3.2　实验要求

要求实验结束时，每位学生均已搭建 HDFS 开发环境；编写了 HDFS 写、读代码；在 client 机上执行了该写、读程序。搭建 HDFS 开发环境，编程实现读写 HDFS，了解 HDFS 读写文件的调用流程，理解 HDFS 读写文件的原理。

3.3　实验原理

3.3.1　Java Classpath

Classpath 设置的目的，在于告诉 Java 执行环境，在哪些目录下可以找到您所要执行的 Java 程序所需要的类或者包。

Java 执行环境本身就是一个平台，执行于这个平台上的程序是已编译完成的 Java 程序（后面会介绍到 Java 程序编译完成之后，会以.class 文件存在）。如果将 Java 执行环境比喻为操作系统，如果设置 Path 变量是为了让操作系统找到指定的工具程序（以 Windows 来说就是找到.exe 文件），则设置 Classpath 的目的就是让 Java 执行环境找到指定的 Java 程序（也就是.class 文件）。

有几个方法可以设置 Classpath，最简单的方法是在系统变量中新增 Classpath 环境变量。以 Windows 7 操作系统为例，右键点击计算机→属性→高级系统设置→环境变量，在弹出菜单的“系统变量”下单击“新建”按钮，在“变量名”文本框中输入 Classpath，在“变量值”文本框中输入 Java 类文件的位置。例如可以输入“；

D:\Java\jdk1.7.0_79\lib\tools.jar; D:\Java\jdk1.7.0_79\lib\rt.jar”，每一路径中间必须以英文;作为分隔，如图 3-1 所示。

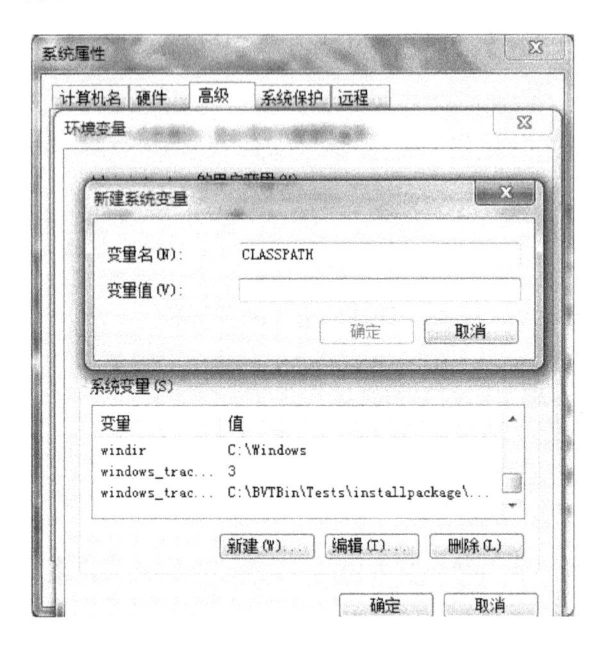

图 3-1 Windows 7 配置 Classpath

事实上 JDK 7.0 默认到当前工作目录（上面的.设置）以及 JDK 的 lib 目录（这里假设是 D:\Java\jdk1.7.0_796\lib）中寻找 Java 程序。所以如果 Java 程序是在这两个目录中，则不必设置 Classpath 变量也可以找得到，将来如果 Java 程序不是放置在这两个目录时，则可以按上述设置 Classpath。

如果所使用的 JDK 工具程序具有 Classpath 命令选项，则可以在执行工具程序时一并指定 Classpath。例如：

javac -classpath classpath1;classpath2...其中 classpath1、classpath 2 是实际要指定的路径。也可以在命令符模式下执行以下的命令，直接设置环境变量，包括 Classpath 变量（这个设置在下次重新打开命令符模式时就不再有效）：

set CLASSPATH=%CLASSPATH%;classpath1;classpath2...总而言之，设置 Classpath 的目的，在于告诉 Java 执行环境，在哪些目录下可以找到您所要执行的 Java 程序（.class 文件）。

3.3.2 Eclipse Hadoop 插件

Eclipse 是一个跨平台的自由集成开发环境（IDE）。通过安装不同的插件，Eclipse 可以支持不同的计算机语言，比如 C++和 Python 等开发工具，亦可以通过 Hadoop 插件来扩展开发 Hadoop 相关程序。

实际工作中，Eclipse Hadoop 插件需要根据 Hadoop 集群的版本号进行下载并编译，过程较为烦琐。为了节约时间，将更多的精力用于实现读写 HDFS 文件，在大数据实验一体机的相关下载页面中已经提供了 2.7.1 版本的 Hadoop 插件和相关的 hadoop 包下载，实验人员可以直接下载这些插件，快速在 Eclipse 中进行安装，开发自己的 Hadoop 程序。

3.4　实验步骤

3.4.1　配置 client 服务器 classpath

使用 SSH 工具登录 client 服务器，执行命令 vi /etc/profile，编辑该文件，将末尾的如下几行：

```
JAVA_HOME=/usr/local/jdk1.7.0_79/
export JRE_HOME=/usr/local/jdk1.7.0_79//jre
export PATH=$PATH:$JAVA_HOME/bin:$JRE_HOME/bin
export CLASSPATH=.:$JAVA_HOME/lib:$JRE_HOME/lib
export HADOOP_HOME=/usr/cstor/hadoop
export PATH=$PATH:$HADOOP_HOME/bin
export HADOOP_COMMON_LIB_NATIVE_DIR=$HADOOP_HOME/lib/native
export HADOOP_OPTS="-Djava.library.path=$HADOOP_HOME/lib"
```

用下列行进行替换：

```
JAVA_HOME=/usr/local/jdk1.7.0_79/
export HADOOP_HOME=/usr/cstor/hadoop
export JRE_HOME=/usr/local/jdk1.7.0_79//jre
export PATH=$PATH:$JAVA_HOME/bin:$JRE_HOME/bin
export
CLASSPATH=.:$JAVA_HOME/lib:$JRE_HOME/lib:$HADOOP_HOME/share/hadoop/common/*:$HADOO
P_HOME/share/hadoop/common/lib/*
export PATH=$PATH:$HADOOP_HOME/bin
export HADOOP_COMMON_LIB_NATIVE_DIR=$HADOOP_HOME/lib/native
export HADOOP_OPTS="-Djava.library.path=$HADOOP_HOME/lib:$HADOOP_HOME/lib/native"
```

执行命令 source /etc/profile，使刚才的环境变量修改生效：

```
[root@client ~]# source /etc/profile
```

3.4.2　在 client 服务器编写 HDFS 写程序

在 client 服务器上执行命令 vi WriteFile.java，编写 HDFS 写文件程序：

```
import org.apache.hadoop.conf.Configuration;
import org.apache.hadoop.fs.FSDataOutputStream;
import org.apache.hadoop.fs.FileSystem;
import org.apache.hadoop.fs.Path;
public class WriteFile {
  public static void main(String[] args)throws Exception{
    Configuration conf=new Configuration();
    FileSystem hdfs = FileSystem.get(conf);
```

```
        Path dfs = new Path("/weather.txt");
        FSDataOutputStream outputStream = hdfs.create(dfs);
        outputStream.writeUTF("nj 20161009 23\n");
        outputStream.close();
    }
}
```

3.4.3 编译并打包 HDFS 写程序

使用 Javac 编译刚刚编写的代码，并使用 jar 命令打包为 hdpAction.jar：

```
[root@client ~]# javac WriteFile.java
[root@client ~]# jar -cvf hdpAction.jar WriteFile.class
added manifest
adding: WriteFile.class(in = 833) (out= 489)(deflated 41%)
```

3.4.4 执行 HDFS 写程序

在 client 服务器上使用 hadoop jar 命令执行 hdpAction.jar：

```
[root@client ~]# hadoop jar  ~/hdpAction.jar  WriteFile
```

查看是否已生成 weather.txt 文件，若已生成，则查看文件内容是否正确：

```
[root@client ~]# hadoop fs -ls /
Found 2 items
-rw-r--r--   3 root supergroup          29 2016-12-05 12:28 /machines
-rw-r--r--   3 root supergroup          17 2016-12-05 14:54 /weather.txt
[root@client ~]# hadoop fs -cat /weather.txt
nj 20161009 23
```

3.4.5 在 client 服务器编写 HDFS 读程序

在 client 服务器上执行命令 vi ReadFile.java，编写 HDFS 读文件程序：

```
import java.io.IOException;

import org.apache.Hadoop.conf.Configuration;
import org.apache.Hadoop.fs.FSDataInputStream;
import org.apache.Hadoop.fs.FileSystem;
import org.apache.Hadoop.fs.Path;

public class ReadFile {
    public static void main(String[] args) throws IOException {
        Configuration conf = new Configuration();
        Path inFile = new Path("/weather.txt");
        FileSystem hdfs = FileSystem.get(conf);
        FSDataInputStream inputStream = hdfs.open(inFile);
        System.out.println("myfile: " + inputStream.readUTF());
```

```
        inputStream.close();
    }
}
```

3.4.6　编译并打包 HDFS 读程序

使用 javac 编译刚刚编写的代码，并使用 jar 命令打包为 hdpAction.jar。

```
[root@client ~]# javac ReadFile.java
[root@client ~]# jar -cvf hdpAction.jar ReadFile.class
added manifest
adding: ReadFile.class(in = 1093) (out= 597)(deflated 45%)
```

3.4.7　执行 HDFS 读程序

在 client 服务器上使用 hadoop jar 命令执行 hdpAction.jar，查看程序运行结果：

```
[root@client ~]# hadoop jar   ~/hdpAction.jar   ReadFile
myfile: nj 20161009 23

[root@client ~]#
```

3.4.8　安装与配置 Eclipse Hadoop 插件

关闭 Eclipse 软件，将 hadoop-eclipse-plugin-2.7.1.jar 文件复制至 Eclipse 安装目录的 plugins 文件夹下，如图 3-2 和图 3-3 所示。

图 3-2　Eclipse 软件的 plugins 文件夹

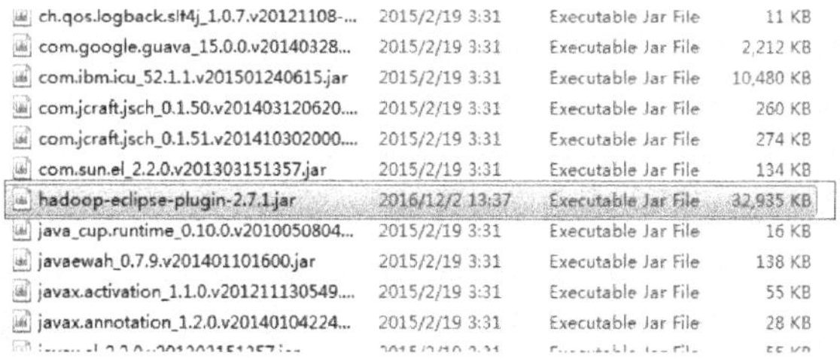

图 3-3　将 hadoop-eclipse-plugin-2.7.1.jar 文件复制至 plugins 文件夹中

接下来，我们需要准备本地的 Hadoop 环境，用于加载 Hadoop 目录中的 jar 包，只需解压 hadoop-2.7.1.tar.gz 文件，解压过程中可能会遇到如下错误，点击关闭忽略即可，如图 3-4 所示。

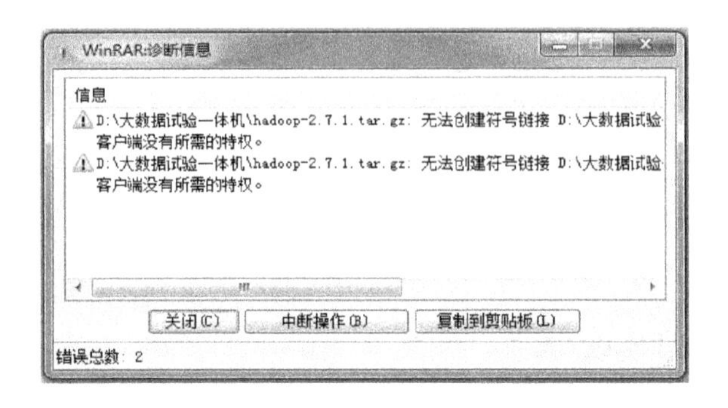

图 3-4　解压 hadoop-2.7.1.tar.gz 可能遇到的错误

现在，我们需要验证是否可以用 Eclipse 新建 Hadoop（HDFS）项目。打开 Eclipse 软件，依次点击 File→New→Other，查看是否已经有 Map/Reduce Project 的选项。第一次新建 Map/Reduce 项目时，需要指定 Hadoop 解压后的位置，具体如图 3-5～图 3-7 所示。

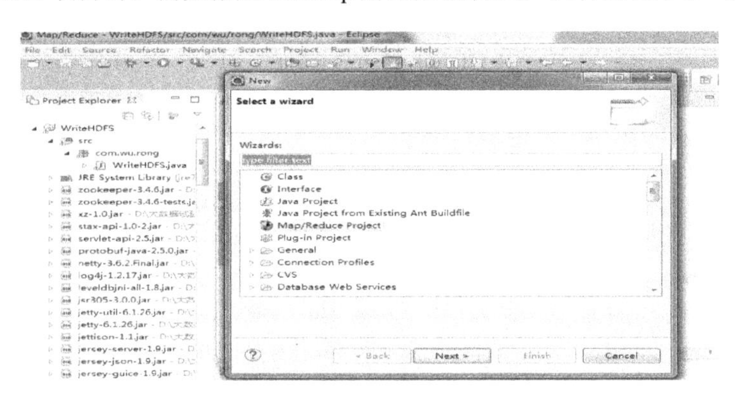

图 3-5　Eclipse 新建 Map/Reduce 项目

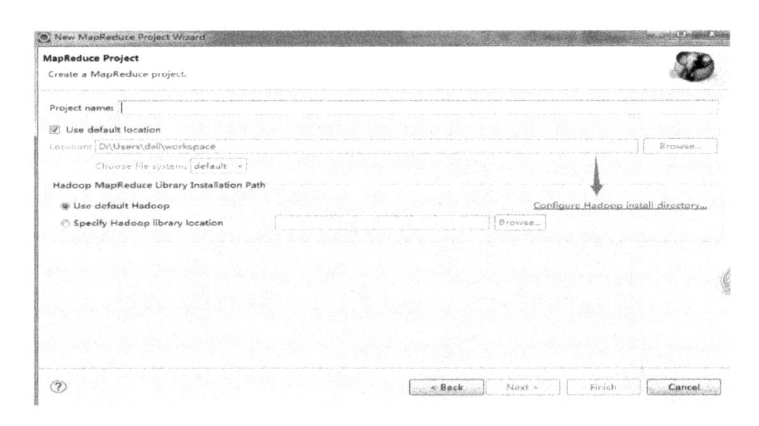

图 3-6　设置 Hadoop 安装目录

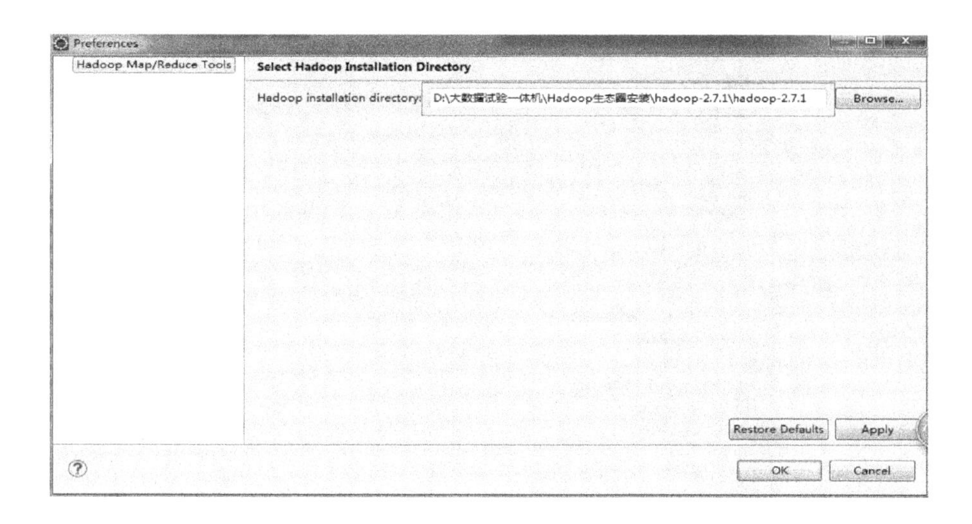

图 3-7 指定 Hadoop 安装目录

3.4.9 使用 Eclipse 开发并打包 HDFS 写文件程序

打开 Eclipse，依次点击 File→New→Map/Reduce Project 或 File→New→Other→Map/Reduce Project，新建项目名为 WriteHDFS 的 Map/Reduce 项目。

新建 WriteFile 类并编写如下代码：

```
import org.apache.hadoop.conf.Configuration;
import org.apache.hadoop.fs.FSDataOutputStream;
import org.apache.hadoop.fs.FileSystem;
import org.apache.hadoop.fs.Path;
public class WriteFile {
  public static void main(String[] args)throws Exception{
    Configuration conf=new Configuration();
    FileSystem hdfs = FileSystem.get(conf);
    Path dfs = new Path("/weather.txt");
    FSDataOutputStream outputStream = hdfs.create(dfs);
    outputStream.writeUTF("nj 20161009 23\n");
    outputStream.close();
  }
}
```

在 Eclipse 左侧的导航栏选中该项目，点击 Export→Java→JAR file，填写导出文件的路径和文件名（本例中设置为 hdpAction.jar），确定导出即可。如图 3-8 和图 3-9 所示。

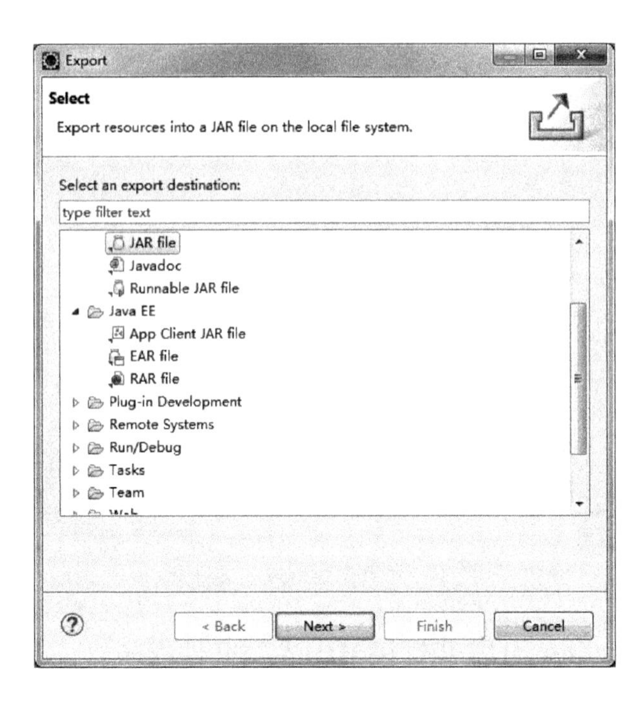

图 3-8　选择导出 JAR 包文件

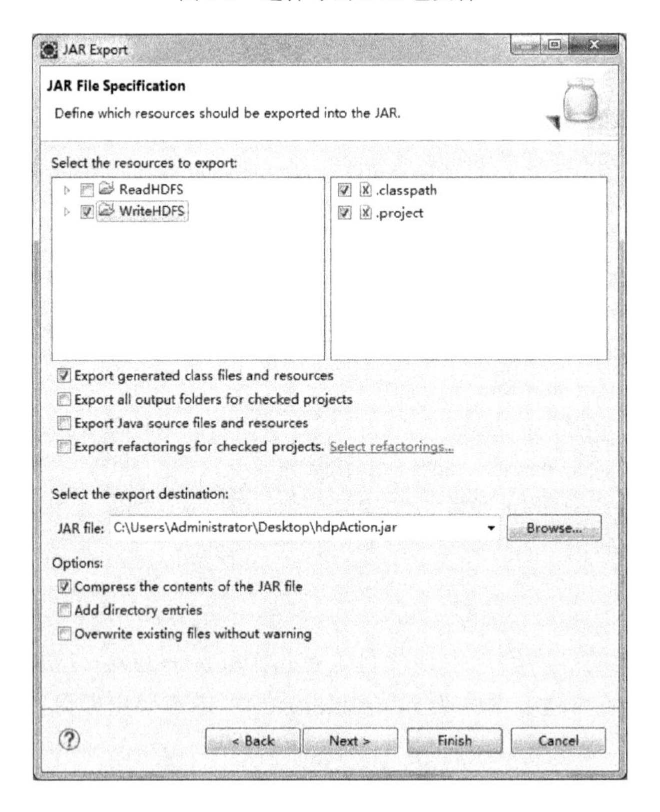

图 3-9　指定导出的 JAR 包文件名

3.4.10 上传HDFS写文件程序jar包并执行

使用WinSCP、XManager或其他SSH工具的sftp工具上传刚刚生成的hdpAction.jar包至client服务器：

```
sftp> lcd C:/Users/Administrator/Desktop/
sftp> put hdpAction.jar
Uploading hdpAction.jar to /root/hdpAction.jar
  100% 2KB        2KB/s 00:00:00
C:/Users/Administrator/Desktop/hdpAction.jar: 2807 bytes transferred in 0 seconds (2807 bytes/s)
```

在client服务器上使用hadoop jar命令执行hdpAction.jar：

```
[root@client ~]# hadoop jar   ~/hdpAction.jar   WriteFile
```

查看是否已生成weather.txt文件，若已生成，则查看文件内容是否正确：

```
[root@client ~]# hadoop fs -ls /
Found 2 items
-rw-r--r--    3 root supergroup            29 2016-12-05 12:28 /machines
-rw-r--r--    3 root supergroup            17 2016-12-05 14:54 /weather.txt
[root@client ~]# Hadoop fs -cat /weather.txt
nj 20161009 23
```

3.4.11 使用Eclipse开发并打包HDFS读文件程序

打开Eclipse，依次点击File→New→Map/Reduce Project或File→New→Other→Map/Reduce Project，新建项目名为ReadHDFS的Map/Reduce项目。

新建ReadFile类并编写如下代码：

```
import java.io.IOException;

import org.apache.hadoop.conf.Configuration;
import org.apache.hadoop.fs.FSDataInputStream;
import org.apache.hadoop.fs.FileSystem;
import org.apache.hadoop.fs.Path;

public class ReadFile {
  public static void main(String[] args) throws IOException {
    Configuration conf = new Configuration();
    Path inFile = new Path("/weather.txt");
    FileSystem hdfs = FileSystem.get(conf);
    FSDataInputStream inputStream = hdfs.open(inFile);
    System.out.println("myfile: " + inputStream.readUTF());
    inputStream.close();
  }
}
```

在Eclipse左侧的导航栏选中该项目，点击Export→Java→JAR file，导出为hdpAction.jar。

3.4.12 上传 HDFS 读文件程序 jar 包并执行

使用 WinSCP、XManager 或其他 SSH 工具的 sftp 工具上传刚刚生成的 hdpAction.jar 包至 client 服务器，并在 client 服务器上使用 hadoop jar 命令执行 hdpAction.jar，查看程序运行结果：

```
[root@client ~]# hadoop jar  ~/hdpAction.jar  ReadFile
myfile: nj 20161009 23

[root@client ~]#
```

实验四　YARN 实验：部署 YARN 集群

4.1　实验目的

了解什么是 YARN 框架，如何搭建 YARN 分布式集群，并能够使用 YARN 集群提交一些简单的任务，理解 YARN 作为 Hadoop 生态中的资源管理器的意义。

4.2　实验要求

搭建 YARN 集群，并使用 YARN 集群提交简单的任务。观察任务提交的之后的 YARN 的执行过程。

4.3　实验原理

4.3.1　YARN 概述

YARN 是一个资源管理、任务调度的框架，采用 master/slave 架构，主要包含三大模块：ResourceManager（RM）、NodeManager（NM）、ApplicationMaster（AM）。其中，ResourceManager 负责所有资源的监控、分配和管理，运行在主节点；NodeManager 负责每一个节点的维护，运行在从节点；ApplicationMaster 负责每一个具体应用程序的调度和协调，只在有任务正在执行时存在。对于所有的 applications，RM 拥有绝对的控制权和对资源的分配权。而每个 AM 则会和 RM 协商资源，同时和 NodeManager 通信来执行和监控 task。几个模块之间的关系如图 4-1 所示。

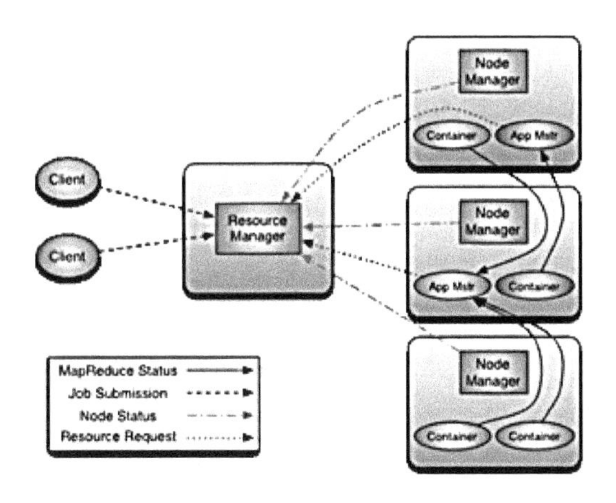

图 4-1　模块间的关系

4.3.2 YARN 运行流程

YARN 运行流程如图 4-2 所示。

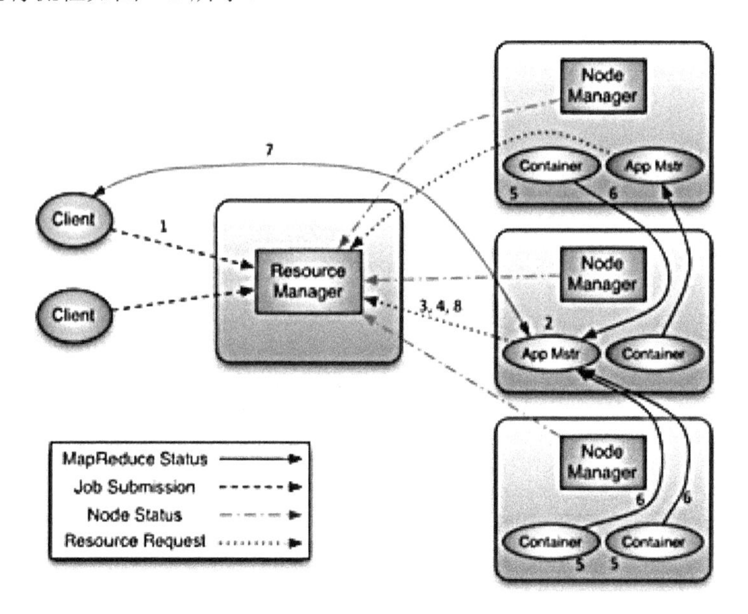

图 4-2　YARN 运行流程

client 向 RM 提交应用程序，其中包括启动该应用的 ApplicationMaster 的必需信息，例如，ApplicationMaster 程序、启动 ApplicationMaster 的命令、用户程序等。

ResourceManager 启动一个 container 用于运行 ApplicationMaster。

启动中的 ApplicationMaster 向 ResourceManager 注册自己，启动成功后与 RM 保持心跳。

ApplicationMaster 向 ResourceManager 发送请求，申请相应数目的 container。

ResourceManager 返回 ApplicationMaster 的申请的 containers 信息。申请成功的 container，由 ApplicationMaster 进行初始化。container 的启动信息初始化后，AM 与对应的 NodeManager 通信，要求 NM 启动 container。AM 与 NM 保持心跳，从而对 NM 上运行的任务进行监控和管理。

container 运行期间，ApplicationMaster 对 container 进行监控。container 通过 RPC 协议向对应的 AM 汇报自己的进度和状态等信息。

应用运行期间，client 直接与 AM 通信获取应用的状态、进度更新等信息。

应用运行结束后，ApplicationMaster 向 ResourceManager 注销自己，并允许属于它的 container 被收回。

4.4 实验步骤

该实验主要分为配置 YARN 的配置文件，启动 YARN 集群，向 YARN 提交几个简单的任务从而了解 YARN 工作的流程。

4.4.1 在 master 机上配置 YARN

操作之前请确认 HDFS 已经启动，具体操作参考之前的实验内容。

指定 YARN 主节点，编辑文件“/usr/cstor/hadoop/etc/hadoop/yarn-site.xml”，将如下内容嵌入此文件里 configuration 标签间：

```
<property><name>yarn.resourcemanager.hostname</name><value>master</value></property>
<property><name>yarn.nodemanager.aux-services</name><value>mapreduce_shuffle</value>
</property>
```

yarn-site.xml 是 YARN 守护进程的配置文件。第一句配置了 ResourceManager 的主机名，第二句配置了节点管理器运行的附加服务为 mapreduce_shuffle，只有这样才可以运行 MapReduce 程序。

在 master 机上操作：将配置好的 YARN 配置文件复制至 slaveX、client。

```
[root@master ~]# cat  ~/data/4/machines
slave1
salve2
slave3
client
[allen@cmaster ~]# for  x  in 'cat ~/data/4/machines' ; do  echo  $x ; scp  /usr/cstor/hadoop/etc/
hadoop/yarn-site.xml  $x:/usr/cstor/hadoop/etc/hadoop/  ; done;
```

4.4.2 统一启动 YARN

确认已配置 slaves 文件，在 master 机器上查看：

```
[root@master ~]# cat  /usr/cstor/hadoop/etc/hadoop/slaves
slave1
slave2
slave3
[root@master ~]#
```

YARN 配置无误，统一启动 YARN：

```
[root@master ~]# /usr/cstor/hadoop/sbin/start-yarn.sh
```

4.4.3 验证 YARN 启动成功

读者可分别在四台机器上执行如下命令，查看 YARN 服务是否已启动。

```
[root@master ~]# jps          #jps 查看 java 进程
```

你会在 master 上看到类似的如下信息：

```
2347  ResourceManager
```

这表明在 master 节点成功启动 ResourceManager，它负责整个集群的资源管理分配，是一个全局的资源管理系统。

而在 slave1、slave2、slave3 上看到类似的如下信息：

```
4021  NodeManager
```

NodeManager 是每个节点上的资源和任务管理器，它是管理这台机器的代理，负责该节点程序的运行，以及该节点资源的管理和监控。YARN 集群每个节点都运行一个 NodeManager。

查看 Web 界面：

在当前的 Windows 机器上打开浏览器，地址栏输入 master 的 IP 和端口号 8088（如 10.1.1.7:8088），可在 Web 界面看到 YARN 相关信息。

4.4.4 在 client 机上提交 DistributedShell 任务

distributedshell，它可以看做 YARN 编程中的“hello world”，它的主要功能是并行执行用户提供的 shell 命令或者 shell 脚本。-jar 指定了包含 ApplicationMaster 的 jar 文件，-shell_command 指定了需要被 ApplicationMaster 执行的 Shell 命令。

在 xshell 上再打开一个 client 的连接，执行：

```
[root@client ~]# /usr/cstor/hadoop/bin/yarn
org.apache.hadoop.yarn.applications.distributedshell.Client  -jar
/usr/cstor/hadoop/share/hadoop/yarn/hadoop-yarn-applications-distributedshell-2.7.1.jar
-shell_command  uptime
```

4.4.5 在 client 机上提交 MapReduce 型任务

（1）指定在 YARN 上运行 MapReduce 任务

首先，在 master 机上，将文件“/usr/cstor/hadoop/etc/hadoop/mapred-site.xml. template”重命名为“/usr/cstor/hadoop/etc/hadoop/mapred-site.xml”。

接着，编辑此文件并将如下内容嵌入此文件的 configuration 标签间：

```
<property><name>mapreduce.framework.name</name><value>yarn</value></property>
```

最后，将 master 机的“/usr/local/hadoop/etc/hadoop/mapred-site.xml”文件复制到 slaveX 与 client，重新启动集群。

（2）在 client 端提交 PI Estimator 任务

首先进入 Hadoop 安装目录：/usr/cstor/hadoop/，然后提交 PI Estimator 任务。

命令最后两个两个参数的含义：第一个参数是指要运行 map 的次数，这里是 2 次；第二个参数是指每个 map 任务，取样的个数；而两数相乘即为总的取样数。PI Estimator 使用 Monte Carlo 方法计算 PI 值的，Monte Carlo 方法可在网络上查找。

```
[root@client hadoop]#  bin/hadoop jar share/hadoop/mapreduce/hadoop-mapreduce-examples-2.7.1.jar
pi 2 10
```

4.5 实验结果

YARN 启动之后在 master 上的 Web 界面上能看到如图 4-3 所示的界面。

All Applications | More protection. The ... | Problem loading page
master:8088/cluster
Search
Logged in as: dr.who
hadoop
All Applications
Cluster
About
Nodes
Node Labels
Applications
NEW
NEW_SAVING
SUBMITTED
ACCEPTED
RUNNING
FINISHED
FAILED
KILLED
Scheduler
Tools

Cluster Metrics

Apps Submitted	Apps Pending	Apps Running	Apps Completed	Containers Running	Memory Used	Memory Total	Memory Reserved	VCores Used	VCores Total	VCores Reserved	Active Nodes	Decommissioned Nodes	Lost Nodes	Unhealthy Nodes	Rebooted Nodes
1	0	0	1	0	0 B	24 GB	0 B	0	24	0	3	0	0	0	0

Scheduler Metrics

Scheduler Type	Scheduling Resource Type	Minimum Allocation	Maximum Allocation
Capacity Scheduler	[MEMORY]	<memory:1024, vCores:1>	<memory:8192, vCores:8>

Show 20 entries　Search:

ID	User	Name	Application Type	Queue	StartTime	FinishTime	State	FinalStatus	Progress	Tracking UI
application_1478175949809_0001	root	DistributedShell	YARN	default	Thu Nov 3 12:35:46 +0000 2016	Thu Nov 3 12:35:54 +0000 2016	FINISHED	SUCCEEDED		History

Showing 1 to 1 of 1 entries　First Previous 1 Next Last

图 4-3　Web 界面总览

提交 DistributedShell 任务之后 Web 界面如图 4-4 所示。

All Applications | More protection. The ... | Problem loading page
master:8088/cluster
Search
Logged in as: dr.who
hadoop
All Applications
Cluster
About
Nodes
Node Labels
Applications
NEW
NEW_SAVING
SUBMITTED
ACCEPTED
RUNNING
FINISHED
FAILED
KILLED
Scheduler
Tools

Cluster Metrics

Apps Submitted	Apps Pending	Apps Running	Apps Completed	Containers Running	Memory Used	Memory Total	Memory Reserved	VCores Used	VCores Total	VCores Reserved	Active Nodes	Decommissioned Nodes	Lost Nodes	Unhealthy Nodes	Rebooted Nodes
1	0	0	1	0	0 B	24 GB	0 B	0	24	0	3	0	0	0	0

Scheduler Metrics

Scheduler Type	Scheduling Resource Type	Minimum Allocation	Maximum Allocation
Capacity Scheduler	[MEMORY]	<memory:1024, vCores:1>	<memory:8192, vCores:8>

Show 20 entries　Search:

ID	User	Name	Application Type	Queue	StartTime	FinishTime	State	FinalStatus	Progress	Tracking UI
application_1478175949809_0001	root	DistributedShell	YARN	default	Thu Nov 3 12:35:46 +0000 2016	Thu Nov 3 12:35:54 +0000 2016	FINISHED	SUCCEEDED		History

Showing 1 to 1 of 1 entries　First Previous 1 Next Last

图 4-4　DistribntedShell 任务提交后的 Web 界面

提交 PI 任务之后 Web 界面如图 4-5 所示。

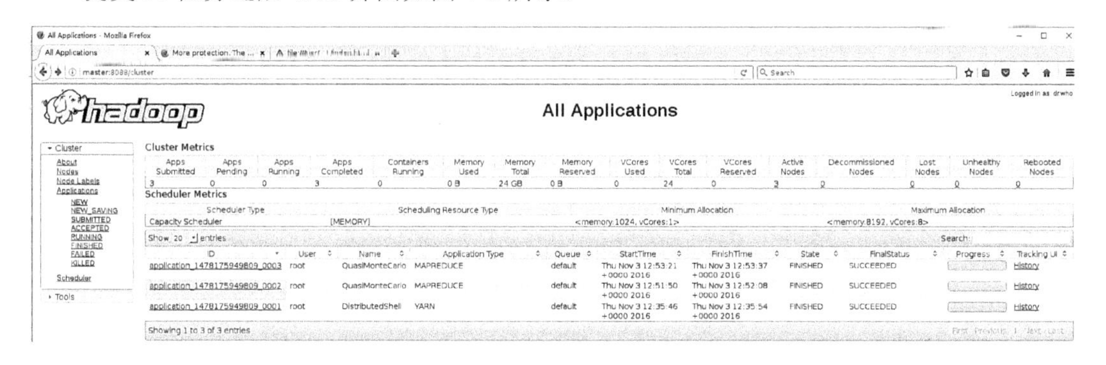

图 4-5　提交 PI 任务后的 Web 界面

在终端能观察到的 MR 任务运行结果如图 4-6 所示。

```
 1 master    2 client    3 client   +
                FILE: Number of write operations=0
                HDFS: Number of bytes read=522
                HDFS: Number of bytes written=215
                HDFS: Number of read operations=11
                HDFS: Number of large read operations=0
                HDFS: Number of write operations=3
        Job Counters
                Launched map tasks=2
                Launched reduce tasks=1
                Data-local map tasks=2
                Total time spent by all maps in occupied slots (ms)=5730
                Total time spent by all reduces in occupied slots (ms)=3177
                Total time spent by all map tasks (ms)=5730
                Total time spent by all reduce tasks (ms)=3177
                Total vcore-seconds taken by all map tasks=5730
                Total vcore-seconds taken by all reduce tasks=3177
                Total megabyte-seconds taken by all map tasks=5867520
                Total megabyte-seconds taken by all reduce tasks=3253248
        Map-Reduce Framework
                Map input records=2
                Map output records=4
                Map output bytes=36
                Map output materialized bytes=56
                Input split bytes=286
                Combine input records=0
                Combine output records=0
                Reduce input groups=2
                Reduce shuffle bytes=56
                Reduce input records=4
                Reduce output records=0
                Spilled Records=8
                Shuffled Maps =2
                Failed Shuffles=0
                Merged Map outputs=2
                GC time elapsed (ms)=99
                CPU time spent (ms)=2640
                Physical memory (bytes) snapshot=704651264
                Virtual memory (bytes) snapshot=2465386496
                Total committed heap usage (bytes)=603979776
        Shuffle Errors
                BAD_ID=0
                CONNECTION=0
                IO_ERROR=0
                WRONG_LENGTH=0
                WRONG_MAP=0
                WRONG_REDUCE=0
        File Input Format Counters
                Bytes Read=236
        File Output Format Counters
                Bytes Written=97
Job Finished in 17.441 seconds
Estimated value of Pi is 3.80000000000000000000
```

图 4-6　MR 任务运行结果

实验五　MapReduce 实验：单词计数

5.1　实验目的

基于 MapReduce 思想，编写 WordCount 程序。

5.2　实验要求

1．理解 MapReduce 编程思想；
2．会编写 MapReduce 版本 WordCount；
3．会执行该程序；
4．自行分析执行过程。

5.3　实验原理

MapReduce 是一种计算模型，简单地说就是将大批量的工作（数据）分解（MAP）执行，然后再将结果合并成最终结果（REDUCE）。这样做的好处是可以在任务被分解后，可以通过大量机器进行并行计算，减少整个操作的时间。

适用范围：数据量大，但是数据种类少可以放入内存。

基本原理及要点：将数据交给不同的机器去处理，数据划分，结果归约。

理解 MapReduce 和 YARN：在新版 Hadoop 中，YARN 作为一个资源管理调度框架，是 Hadoop 下 MapReduce 程序运行的生存环境。其实 MapRuduce 除了可以运行 YARN 框架下，也可以运行在诸如 Mesos，Corona 之类的调度框架上，使用不同的调度框架，需要针对 Hadoop 做不同的适配。

一个完成的 MapReduce 程序在 YARN 中执行过程如下：

（1）ResourcManager JobClient 向 ResourcManager 提交一个 job。

（2）ResourcManager 向 Scheduler 请求一个供 MRAppMaster 运行的 container，然后启动它。

（3）MRAppMaster 启动起来后向 ResourcManager 注册。

（4）ResourcManagerJobClient 向 ResourcManager 获取到 MRAppMaster 相关的信息，然后直接与 MRAppMaster 进行通信。

（5）MRAppMaster 算 splits 并为所有的 map 构造资源请求。

（6）MRAppMaster 做一些必要的 MR OutputCommitter 的准备工作。

（7）MRAppMaster 向 RM（Scheduler）发起资源请求，得到一组供 map/reduce task 运行的 container，然后与 NodeManager 一起对每一个 container 执行一些必要的任务，包括资源本地化等。

（8）MRAppMaster 监视运行着的 task 直到完成，当 task 失败时，申请新的 container 运行失败的 task。

（9）当每个 map/reduce task 完成后，MRAppMaster 运行 MR OutputCommitter 的 cleanup 代码，也就是进行一些收尾工作。

（10）当所有的 map/reduce 完成后，MRAppMaster 运行 OutputCommitter 的必要的 job commit 或者 abort APIs。

（11）MRAppMaster 退出。

5.3.1 MapReduce 编程

编写在 Hadoop 中依赖 YARN 框架执行的 MapReduce 程序，并不需要自己开发 MRAppMaster 和 YARNRunner，因为 Hadoop 已经默认提供通用的 YARNRunner 和 MRAppMaster 程序，大部分情况下只需要编写相应的 Map 处理和 Reduce 处理过程的业务程序即可。

编写一个 MapReduce 程序并不复杂，关键点在于掌握分布式的编程思想和方法，主要将计算过程分为以下五个步骤：

（1）迭代，遍历输入数据，并将之解析成 key/value 对。

（2）将输入 key/value 对映射（map）成另外一些 key/value 对。

（3）依据 key 对中间数据进行分组（grouping）。

（4）以组为单位对数据进行归约（reduce）。

（5）迭代，将最终产生的 key/value 对保存到输出文件中。

5.3.2 Java API 解析

（1）InputFormat：用于描述输入数据的格式，常用的为 TextInputFormat 提供如下两个功能。

数据切分：按照某个策略将输入数据切分成若干个 split，以便确定 Map Task 个数以及对应的 split。

为 Mapper 提供数据：给定某个 split，能将其解析成一个个 key/value 对。

（2）OutputFormat：用于描述输出数据的格式，它能够将用户提供的 key/value 对写入特定格式的文件中。

（3）Mapper/Reducer：Mapper/Reducer 中封装了应用程序的数据处理逻辑。

（4）Writable：Hadoop 自定义的序列化接口。实现该类的接口可以用作 MapReduce 过程中的 value 数据使用。

（5）WritableComparable：在 Writable 基础上继承了 Comparable 接口，实现该类的接口可以用作 MapReduce 过程中的 key 数据使用（因为 key 包含了比较排序的操作）。

5.4　实验步骤

本实验主要分为，确认前期准备，编写 MapReduce 程序，打包提交代码，查看运行结果等步骤，详细如下。

5.4.1　启动 Hadoop

执行命令启动前面实验部署好的 Hadoop 系统。

```
[root@master ~]# cd /usr/cstor/hadoop/
[root@master hadoop]# sbin/start-all.sh
```

5.4.2　验证 HDFS 上没有 WordCount 的文件夹

```
[root@client ~]# cd /usr/cstor/hadoop/
[root@client hadoop]# bin/hadoop fs -ls /                    #查看 HDFS 上根目录文件 /
```

此时 HDFS 上应该是没有 WordCount 文件夹。

5.4.3　上传数据文件到 HDFS

```
[root@client ~]# cd /usr/cstor/hadoop/
[root@client hadoop]# bin/hadoop fs -put /root/data/5/word   /
```

5.4.4　编写 MapReduce 程序

主要编写 Map 和 Reduce 类，其中 Map 过程需要继承 org.apache.hadoop.mapreduce 包中 Mapper 类，并重写其 map 方法；Reduce 过程需要继承 org.apache.hadoop.mapreduce 包中 Reduce 类，并重写其 reduce 方法。

```
import org.apache.hadoop.conf.Configuration;
import org.apache.hadoop.fs.Path;
import org.apache.hadoop.io.IntWritable;
import org.apache.hadoop.io.Text;
import org.apache.hadoop.mapreduce.Job;
import org.apache.hadoop.mapreduce.Mapper;
import org.apache.hadoop.mapreduce.Reducer;
import org.apache.hadoop.mapreduce.lib.input.TextInputFormat;
import org.apache.hadoop.mapreduce.lib.output.TextOutputFormat;
import org.apache.hadoop.mapreduce.lib.partition.HashPartitioner;

  import java.io.IOException;
  import java.util.StringTokenizer;


public class WordCount {
    public static class TokenizerMapper extends Mapper<Object, Text, Text, IntWritable> {
```

```
        private final static IntWritable one = new IntWritable(1);
        private Text word = new Text();
        //map 方法，划分一行文本，读一个单词写出一个<单词,1>
        public   void   map(Object   key,   Text   value,   Context   context)throws   IOException,
InterruptedException {
            StringTokenizer itr = new StringTokenizer(value.toString());
            while (itr.hasMoreTokens()) {
                word.set(itr.nextToken());
                context.write(word, one);//写出<单词,1>
            }}}
    //定义 reduce 类，对相同的单词，把它们<K,VList>中的 VList 值全部相加
    public static class IntSumReducer extends Reducer<Text, IntWritable, Text, IntWritable> {
        private IntWritable result = new IntWritable();
        public void reduce(Text key, Iterable<IntWritable> values,Context context)
                throws IOException, InterruptedException {
            int sum = 0;
            for (IntWritable val : values) {
                sum += val.get();//相当于<Hello,1><Hello,1>，将两个 1 相加
            }
            result.set(sum);
            context.write(key, result);//写出这个单词，和这个单词出现次数<单词，单词出现次数>
        }}
    public static void main(String[] args) throws Exception {//主方法，函数入口
        Configuration conf = new Configuration();                //实例化配置文件类
        Job job = new Job(conf, "WordCount");                    //实例化 Job 类
        job.setInputFormatClass(TextInputFormat.class);          //指定使用默认输入格式类
        TextInputFormat.setInputPaths(job, args[0]);             //设置待处理文件的位置
        job.setJarByClass(WordCount.class);                      //设置主类名
        job.setMapperClass(TokenizerMapper.class);               //指定使用上述自定义 Map 类
        job.setCombinerClass(IntSumReducer.class);               //指定开启 Combiner 函数
        job.setMapOutputKeyClass(Text.class);                    //指定 Map 类输出的<K,V>，K 类型
        job.setMapOutputValueClass(IntWritable.class);           //指定 Map 类输出的<K,V>，V 类型
        job.setPartitionerClass(HashPartitioner.class);          //指定使用默认的 HashPartitioner 类
        job.setReducerClass(IntSumReducer.class);                //指定使用上述自定义 Reduce 类
        job.setNumReduceTasks(Integer.parseInt(args[2]));        //指定 Reduce 个数
        job.setOutputKeyClass(Text.class);                       //指定 Reduce 类输出的<K,V>,K 类型
        job.setOutputValueClass(Text.class);                     //指定 Reduce 类输出的<K,V>,V 类型
        job.setOutputFormatClass(TextOutputFormat.class);        //指定使用默认输出格式类
        TextOutputFormat.setOutputPath(job, new Path(args[1]));  //设置输出结果文件位置
        System.exit(job.waitForCompletion(true) ? 0 : 1);        //提交任务并监控任务状态
    }
}
```

5.4.5 使用 Eclipse 开发工具将该代码打包

假定打包后的文件名为 hdpAction.jar，主类 WordCount 位于包 njupt 下，则可使用如下命令向 YARN 集群提交本应用。

```
[root@client ~]# ./yarn  jar  hdpAction.jar  njupt.WordCount  /word  /wordcount 1
```

其中“yarn”为命令，“jar”为命令参数，后面紧跟打包后的代码地址，“njupt”为包名，“WordCount”为主类名，“/word”为输入文件在 HDFS 中的位置，“/wordcount”为输出文件在 HDFS 中的位置。

5.5 实验结果

5.5.1 程序运行成功控制台上的显示内容

提交 WordCount 如图 5-1 所示。

```
16/11/05 14:58:11 INFO client.RMProxy: Connecting to ResourceManager at master/172.17.0.2:8032
16/11/05 14:58:12 WARN mapreduce.JobResourceUploader: Hadoop command-line option parsing not performed. Implement the Tool interface and execute your application with ToolRunner to remedy this.
16/11/05 14:58:12 INFO input.FileInputFormat: Total input paths to process : 1
16/11/05 14:58:12 INFO mapreduce.JobSubmitter: number of splits:1
16/11/05 14:58:12 INFO mapreduce.JobSubmitter: Submitting tokens for job: job_1478265821675_0004
16/11/05 14:58:12 INFO impl.YarnClientImpl: Submitted application application_1478265821675_0004
16/11/05 14:58:13 INFO mapreduce.Job: The url to track the job: http://master:8088/proxy/application_1478265821675_0004/
16/11/05 14:58:13 INFO mapreduce.Job: Running job: job_1478265821675_0004
16/11/05 14:58:18 INFO mapreduce.Job: Job job_1478265821675_0004 running in uber mode : false
16/11/05 14:58:18 INFO mapreduce.Job:  map 0% reduce 0%
16/11/05 14:58:23 INFO mapreduce.Job:  map 100% reduce 0%
16/11/05 14:58:28 INFO mapreduce.Job:  map 100% reduce 75%
16/11/05 14:58:29 INFO mapreduce.Job:  map 100% reduce 100%
16/11/05 14:58:29 INFO mapreduce.Job: Job job_1478265821675_0004 completed successfully
16/11/05 14:58:29 INFO mapreduce.Job: Counters: 49
	File System Counters
		FILE: Number of bytes read=78
		FILE: Number of bytes written=580273
		FILE: Number of read operations=0
		FILE: Number of large read operations=0
		FILE: Number of write operations=0
		HDFS: Number of bytes read=124
		HDFS: Number of bytes written=34
		HDFS: Number of read operations=15
		HDFS: Number of large read operations=0
		HDFS: Number of write operations=8
	Job Counters
		Launched map tasks=1
		Launched reduce tasks=4
		Data-local map tasks=1
		Total time spent by all maps in occupied slots (ms)=2508
		Total time spent by all reduces in occupied slots (ms)=11972
		Total time spent by all map tasks (ms)=2508
		Total time spent by all reduce tasks (ms)=11972
		Total vcore-seconds taken by all map tasks=2508
		Total vcore-seconds taken by all reduce tasks=11972
		Total megabyte-seconds taken by all map tasks=2568192
		Total megabyte-seconds taken by all reduce tasks=12259328
	Map-Reduce Framework
		Map input records=2
		Map output records=7
		Map output bytes=62
		Map output materialized bytes=78
		Input split bytes=88
		Combine input records=7
		Combine output records=5
		Reduce input groups=5
		Reduce shuffle bytes=78
		Reduce input records=5
		Reduce output records=5
		Spilled Records=10
		Shuffled Maps =4
		Failed Shuffles=0
		Merged Map outputs=4
		GC time elapsed (ms)=159
		CPU time spent (ms)=5140
		Physical memory (bytes) snapshot=934477824
		Virtual memory (bytes) snapshot=4098646016
		Total committed heap usage (bytes)=1006632960
	Shuffle Errors
		BAD_ID=0
		CONNECTION=0
		IO_ERROR=0
		WRONG_LENGTH=0
		WRONG_MAP=0
		WRONG_REDUCE=0
	File Input Format Counters
		Bytes Read=36
	File Output Format Counters
		Bytes Written=34
[root@client hadoop]# ~
```

图 5-1 提交 WordCount

5.5.2 在 HDFS 上查看结果

查看 WordCount 结果如图 5-2 所示。

```
[root@client hadoop]# bin/hadoop fs -ls /wordcount
16/11/05 15:10:35 WARN util.NativeCodeLoader: Unable to load native-hadoop library for your platform... using builtin-java classes where applicable
Found 2 items
-rw-r--r--   3 root supergroup          0 2016-11-05 15:09 /wordcount/_SUCCESS
-rw-r--r--   3 root supergroup         52 2016-11-05 15:09 /wordcount/part-r-00000
[root@client hadoop]# bin/hadoop fs -cat /wordcount/part-r-00000
16/11/05 15:10:38 WARN util.NativeCodeLoader: Unable to load native-hadoop library for your platform... using builtin-java classes where applicable
aaa     5
aaaaaaaa        1
bbb     1
cccccc  1
ddddddd 1
wwwwwww 1
[root@client hadoop]#
```

图 5-2 查看 WordCount 结果

实验六　MapReduce 实验：二次排序

6.1　实验目的

基于 MapReduce 思想，编写 SecondarySort 程序。

6.2　实验要求

要能理解 MapReduce 编程思想，会编写 MapReduce 版本二次排序程序，然后将其执行并分析执行过程。

6.3　实验原理

MR 默认会对键进行排序，然而有的时候我们也有对值进行排序的需求。满足这种需求一是可以在 reduce 阶段排序收集过来的 values，但是，如果有数量巨大的 values 可能就会导致内存溢出等问题，这就是二次排序应用的场景——将对值的排序也安排到 MR 计算过程之中，而不是单独来做。

二次排序就是首先按照第一字段排序，然后再对第一字段相同的行按照第二字段排序，注意不能破坏第一次排序的结果。

6.4　实验步骤

6.4.1　编写程序

程序主要难点在于排序和聚合。

对于排序我们需要定义一个 IntPair 类用于数据的存储，并在 IntPair 类内部自定义 Comparator 类以实现第一字段和第二字段的比较。

对于聚合我们需要定义一个 FirstPartitioner 类，在 FirstPartitioner 类内部指定聚合规则为第一字段。

此外，我们还需要开启 MapReduce 框架自定义 Partitioner 功能和 GroupingComparator 功能。

IntPair 类：

```
package mr;

import java.io.DataInput;
import java.io.DataOutput;
import java.io.IOException;

import org.apache.hadoop.io.IntWritable;
import org.apache.hadoop.io.WritableComparable;

public class IntPair implements WritableComparable<IntPair> {
    private IntWritable first;
    private IntWritable second;
    public void set(IntWritable first, IntWritable second) {
        this.first = first;
        this.second = second;
    }
    //注意：需要添加无参的构造方法，否则反射时会报错。
    public IntPair() {
        set(new IntWritable(), new IntWritable());
    }
    public IntPair(int first, int second) {
        set(new IntWritable(first), new IntWritable(second));
    }
    public IntPair(IntWritable first, IntWritable second) {
        set(first, second);
    }
    public IntWritable getFirst() {
        return first;
    }
    public void setFirst(IntWritable first) {
        this.first = first;
    }
    public IntWritable getSecond() {
        return second;
    }
    public void setSecond(IntWritable second) {
        this.second = second;
    }
    public void write(DataOutput out) throws IOException {
        first.write(out);
        second.write(out);
    }
    public void readFields(DataInput in) throws IOException {
        first.readFields(in);
```

```
            second.readFields(in);
        }
        public int hashCode() {
            return first.hashCode() * 163 + second.hashCode();
        }
        public boolean equals(Object o) {
            if (o instanceof IntPair) {
                IntPair tp = (IntPair) o;
                return first.equals(tp.first) && second.equals(tp.second);
            }
            return false;
        }
        public String toString() {
            return first + "\t" + second;
        }
        public int compareTo(IntPair tp) {
            int cmp = first.compareTo(tp.first);
            if (cmp != 0) {
                return cmp;
            }
            return second.compareTo(tp.second);
        }
    }
```

完整代码：

```
package mr;

import java.io.IOException;

import org.apache.hadoop.conf.Configuration;
import org.apache.hadoop.fs.Path;
import org.apache.hadoop.io.LongWritable;
import org.apache.hadoop.io.NullWritable;
import org.apache.hadoop.io.Text;
import org.apache.hadoop.io.WritableComparable;
import org.apache.hadoop.io.WritableComparator;
import org.apache.hadoop.mapreduce.Job;
import org.apache.hadoop.mapreduce.Mapper;
import org.apache.hadoop.mapreduce.Partitioner;
import org.apache.hadoop.mapreduce.Reducer;
import org.apache.hadoop.mapreduce.lib.input.FileInputFormat;
import org.apache.hadoop.mapreduce.lib.output.FileOutputFormat;

public class SecondarySort {
    static class TheMapper extends Mapper<LongWritable, Text, IntPair, NullWritable> {
```

```
        @Override
        protected void map(LongWritable key, Text value, Context context)
                throws IOException, InterruptedException {
            String[] fields = value.toString().split("\t");
            int field1 = Integer.parseInt(fields[0]);
            int field2 = Integer.parseInt(fields[1]);
            context.write(new IntPair(field1,field2), NullWritable.get());
        }
    }
    static class TheReducer extends Reducer<IntPair, NullWritable,IntPair, NullWritable> {
        //private static final Text SEPARATOR = new Text("------------------------------------------------");
        @Override
        protected void reduce(IntPair key, Iterable<NullWritable> values, Context context)
                throws IOException, InterruptedException {
            context.write(key, NullWritable.get());
        }
    }
    public static class FirstPartitioner extends Partitioner<IntPair, NullWritable> {
        public int getPartition(IntPair key, NullWritable value,
                int numPartitions) {
            return Math.abs(key.getFirst().get()) % numPartitions;
        }
    }
    //如果不添加这个类，默认第一列和第二列都是升序排序的。
    //这个类的作用是使第一列升序排序，第二列降序排序
    public static class KeyComparator extends WritableComparator {
        //无参构造器必须加上，否则报错。
        protected KeyComparator() {
            super(IntPair.class, true);
        }
        public int compare(WritableComparable a, WritableComparable b) {
            IntPair ip1 = (IntPair) a;
            IntPair ip2 = (IntPair) b;
            //第一列按升序排序
            int cmp = ip1.getFirst().compareTo(ip2.getFirst());
            if (cmp != 0) {
                return cmp;
            }
            //在第一列相等的情况下，第二列按倒序排序
            return -ip1.getSecond().compareTo(ip2.getSecond());
        }
    }
    //入口程序
    public static void main(String[] args) throws Exception {
```

```
        Configuration conf = new Configuration();
        Job job = Job.getInstance(conf);
        job.setJarByClass(SecondarySort.class);
        //设置 Mapper 的相关属性
        job.setMapperClass(TheMapper.class);
        //当 Mapper 中的输出的 key 和 value 的类型和 Reduce 输出
        //的 key 和 value 的类型相同时，以下两句可以省略。
        //job.setMapOutputKeyClass(IntPair.class);
        //job.setMapOutputValueClass(NullWritable.class);
        FileInputFormat.setInputPaths(job, new Path(args[0]));
        //设置分区的相关属性
        job.setPartitionerClass(FirstPartitioner.class);
        //在 map 中对 key 进行排序
        job.setSortComparatorClass(KeyComparator.class);
        //job.setGroupingComparatorClass(GroupComparator.class);
        //设置 Reducer 的相关属性
        job.setReducerClass(TheReducer.class);
        job.setOutputKeyClass(IntPair.class);
        job.setOutputValueClass(NullWritable.class);
        FileOutputFormat.setOutputPath(job, new Path(args[1]));
        //设置 Reducer 数量
        int reduceNum = 1;
        if(args.length >= 3 && args[2] != null){
            reduceNum = Integer.parseInt(args[2]);
        }
        job.setNumReduceTasks(reduceNum);
        job.waitForCompletion(true);
    }
}
```

6.4.2 打包提交

使用 Eclipse 开发工具将该代码打包，选择主类为 mr.Secondary。如果没有指定主类，那么在执行时就要指定须执行的类。假定打包后的文件名为 Secondary.jar，主类 SecondarySort 位于包 mr 下，则可使用如下命令向 Hadoop 集群提交本应用。

```
[root@client hadoop]# bin/hadoop jar SecondarySort.jar mr.Secondary /user/mapreduce/secsort/in/secsortdata.txt  /user/mapreduce/secsort/out  1
```

其中“hadoop”为命令，“jar”为命令参数，后面紧跟打的包，/user/mapreduce/secsort/in/secsortdata.txt”为输入文件在 HDFS 中的位置，如果 HDFS 中没有这个文件，则自己自行上传。“/user/mapreduce/secsort/out/”为输出文件在 HDFS 中的位置，“1”为 Reduce 个数。

6.5 实验结果

6.5.1 输入数据

输入数据如下：secsortdata.txt ('/t'分割)（数据放在/root/data/6 目录下）。

```
7	444
3	9999
7	333
4	22
3	7777
7	555
3	6666
6	0
3	8888
4	11
```

6.5.2 执行结果

在 client 上执行对 HDFS 上的文件/user/mapreduce/secsort/out/part-r-00000 内容查看的操作：

```
[root@client hadoop]# bin/hadoop fs -cat  /user/mapreduce/secsort/out/p*
```

查看二次排序结果如图 6-1 所示。

```
[root@client hadoop]# bin/hadoop fs -cat  /user/mapreduce/secsort/out/p*
16/11/07 16:12:14 WARN util.NativeCodeLoader: Unable to load native-hadoop library f
e applicable
3	9999
3	8888
3	7777
3	6666
4	22
4	11
6	0
7	555
7	444
7	333
[root@client hadoop]#
```

图 6-1　查看二次排序结果

实验七　MapReduce 实验：计数器

7.1　实验目的

基于 MapReduce 思想，编写计数器程序。

7.2　实验要求

能够理解 MapReduce 编程思想，然后会编写 MapReduce 版本计数器程序，并能执行该程序和分析执行过程。

7.3　实验背景

7.3.1　MapReduce 计数器是什么?

计数器用来记录 job 的执行进度和状态。它的作用可以理解为日志，我们可以在程序的某个位置插入计数器，记录数据或者进度的变化情况。

7.3.2　MapReduce 计数器能做什么?

MapReduce 计数器（Counter）为我们提供一个窗口，用于观察 MapReduce Job 运行期的各种细节数据。对 MapReduce 性能调优很有帮助，MapReduce 性能优化的评估大部分都是基于这些 Counter 的数值表现出来的。

在许多情况下，一个用户需要了解待分析的数据，尽管这并非所要执行的分析任务的核心内容。以统计数据集中无效记录数目的任务为例，如果发现无效记录的比例相当高，那么就需要认真思考为何存在如此多无效记录。是所采用的检测程序存在缺陷，还是数据集质量确实很低，包含大量无效记录？如果确定是数据集的质量问题，则可能需要扩大数据集的规模，以增大有效记录的比例，从而进行有意义的分析。

计数器是一种收集作业统计信息的有效手段，用于质量控制或应用级统计。计数器还可辅助诊断系统故障。如果需要将日志信息传输到 map 或 reduce 任务，更好的方法通常是尝试传输计数器值以监测某一特定事件是否发生。对于大型分布式作业而言，使用计数器更为方便。首先，获取计数器值比输出日志更方便，其次，根据计数器值统计特定事件的发生次数要比分析一堆日志文件容易得多。

7.3.3 内置计数器

MapReduce 自带了许多默认 Counter，现在我们来分析这些默认 Counter 的含义，方便大家观察 Job 结果，如输入的字节数、输出的字节数、Map 端输入/输出的字节数和条数、Reduce 端的输入/输出的字节数和条数等。下面我们只需了解这些内置计数器，知道计数器组名称（GroupName）和计数器名称（CounterName），以后使用计数器会查找 GroupName 和 CounterName 即可。

7.3.4 计数器使用

1. 定义计数器

枚举声明计数器：

```
// 自定义枚举变量 Enum
Counter counter = context.getCounter(Enum enum)
```

自定义计数器：

```
// 自己命名 groupName 和 counterName
Counter counter = context.getCounter(String groupName,String counterName)
```

2. 为计数器赋值

初始化计数器：

```
counter.setValue(long value);//设置初始值
```

计数器自增：

```
counter.increment(long incr);// 增加计数
```

3. 获取计数器的值

获取枚举计数器的值：

```
Configuration conf = new Configuration();
Job job = new Job(conf, "MyCounter");
job.waitForCompletion(true);
Counters counters=job.getCounters();
Counter counter=counters.findCounter(LOG_PROCESSOR_COUNTER.BAD_RECORDS_LONG);//查找枚举计数器，假如 Enum 的变量为 BAD_RECORDS_LONG
long value=counter.getValue();//获取计数值
```

获取自定义计数器的值：

```
Configuration conf = new Configuration();
Job job = new Job(conf, "MyCounter");
job.waitForCompletion(true);
Counters counters = job.getCounters();
Counter counter=counters.findCounter("ErrorCounter","toolong");// 假如 groupName 为 ErrorCounter，counterName 为 toolong
long value = counter.getValue();// 获取计数值
```

获取内置计数器的值：

```
Configuration conf = new Configuration();
Job job = new Job(conf, "MyCounter");
job.waitForCompletion(true);
Counters counters=job.getCounters();
// 查找作业运行启动的 reduce 个数的计数器，groupName 和 counterName 可以从内置计数器表格查询（前面已经列举有）
Counter counter=counters.findCounter("org.apache.hadoop.mapreduce.JobCounter","TOTAL_LAUNCHED_REDUCES") ;// 假如 groupName 为 org.apache.hadoop.mapreduce.JobCounter ， counterName 为 TOTAL_LAUNCHED_REDUCES
long value=counter.getValue();// 获取计数值
```

获取所有计数器的值：

```
Configuration conf = new Configuration();
Job job = new Job(conf, "MyCounter");
Counters counters = job.getCounters();
for (CounterGroup group : counters) {
  for (Counter counter : group) {
    System.out.println(counter.getDisplayName() + ": " + counter.getName() + ": "+ counter.getValue());
  }
}
```

7.3.5 自定义计数器

MapReduce 允许用户编写程序来定义计数器，计数器的值可在 Mapper 或 Reducer 中增加。多个计数器由一个 Java 枚举（enum）类型来定义，以便对计数器分组。一个作业可以定义的枚举类型数量不限，各个枚举类型所包含的字段数量也不限。枚举类型的名称即为组的名称，枚举类型的字段就是计数器名称。计数器是全局的。换言之，MapReduce 框架将跨所有 map 和 reduce 聚集这些计数器，并在作业结束时产生一个最终结果。

7.4 实验步骤

7.4.1 实验分析设计

该实验要求学生自己实现一个计数器，统计输入的无效数据。说明如下：

假如一个文件，规范的格式是 3 个字段，“\t” 作为分隔符，其中有 2 条异常数据，一条数据是只有 2 个字段，一条数据是有 4 个字段。其内容如下所示：

```
jim	1	28
kate	0	26
tom	1
lily	0	29	22
```

编写代码统计文档中字段不为 3 个的异常数据个数。如果字段超过 3 个视为过长字段，字段少于 3 个视为过短字段。

7.4.2 编写程序

完整代码：

```
package mr ;
import java.io.IOException;
import org.apache.hadoop.conf.Configuration;
import org.apache.hadoop.fs.Path;
import org.apache.hadoop.io.LongWritable;
import org.apache.hadoop.io.Text;
import org.apache.hadoop.mapreduce.Counter;
import org.apache.hadoop.mapreduce.Job;
import org.apache.hadoop.mapreduce.Mapper;
import org.apache.hadoop.mapreduce.lib.input.FileInputFormat;
import org.apache.hadoop.mapreduce.lib.output.FileOutputFormat;
import org.apache.hadoop.util.GenericOptionsParser;

public class Counters {
 public static class MyCounterMap extends Mapper<LongWritable, Text, Text, Text> {
        public static Counter ct = null;
        protected void map(LongWritable key, Text value,
                    org.apache.hadoop.mapreduce.Mapper<LongWritable,  Text,  Text,  Text>.Context
context)
                    throws java.io.IOException, InterruptedException {
              String arr_value[] = value.toString().split("\t");
 if (arr_value.length > 3) {
 ct = context.getCounter("ErrorCounter", "toolong"); // ErrorCounter 为组名，toolong 为组员名
ct.increment(1); //  计数器加一
} else if (arr_value.length < 3) {
 ct = context.getCounter("ErrorCounter", "tooshort");
              ct.increment(1);
        }
 }
}
public  static  void  main(String[]  args)  throws  IOException,  InterruptedException,
ClassNotFoundException {
 Configuration conf = new Configuration();
```

```
    String[] otherArgs = new GenericOptionsParser(conf, args).getRemainingArgs();
    if (otherArgs.length != 2) {
            System.err.println("Usage: Counters <in> <out>");
    System.exit(2);
   }
   Job job = new Job(conf, "Counter");
            job.setJarByClass(Counters.class);

            job.setMapperClass(MyCounterMap.class);

            FileInputFormat.addInputPath(job, new Path(otherArgs[0]));
            FileOutputFormat.setOutputPath(job, new Path(otherArgs[1]));
            System.exit(job.waitForCompletion(true) ? 0 : 1);
    }
}
```

7.4.3　打包并提交

使用 Eclipse 开发工具将该代码打包，选择主类为 mr.Counters。假定打包后的文件名为 Counters.jar，主类 Counters 位于包 mr 下，则可使用如下命令向 Hadoop 集群提交本应用。

```
[root@client hadoop]# bin/hadoop jar Counters.jar mr.Counters /usr/counters/in/counters.txt /usr/counters/out
```

其中“hadoop”为命令，“jar”为命令参数，后面紧跟打包。“/usr/counts/in/counts.txt”为输入文件在 HDFS 中的位置（如果没有，自行上传），“/usr/counts/out”为输出文件在 HDFS 中的位置。

7.5　实验结果

7.5.1　输入数据

输入数据如下：counters.txt　（/t 分割）（数据统一放在/root/data 目录下）。

```
jim     1    28
kate    0    26
tom     1
lily    0    29    22
```

7.5.2 输出显示

提交计数器输出如图 7-1 所示。

```
                Reduce shuffle bytes=6
                Reduce input records=0
                Reduce output records=0
                Spilled Records=0
                Shuffled Maps =1
                Failed Shuffles=0
                Merged Map outputs=1
                GC time elapsed (ms)=68
                CPU time spent (ms)=1590
                Physical memory (bytes) snapshot=426364928
                Virtual memory (bytes) snapshot=1632751616
                Total committed heap usage (bytes)=402653184
        ErrorCounter
                toolong=1
                tooshort=1
        Shuffle Errors
                BAD_ID=0
                CONNECTION=0
                IO_ERROR=0
                WRONG_LENGTH=0
                WRONG_MAP=0
                WRONG_REDUCE=0
        File Input Format Counters
                Bytes Read=38
        File Output Format Counters
                Bytes Written=0
[root@client hadoop]#
```

图 7-1　提交计数器输出

实验八　MapReduce 实验：Join 操作

8.1　实验目的

基于 MapReduce 思想，编写两文件 Join 操作的程序。

8.2　实验要求

能够理解 MapReduce 编程思想，然后会编写 MapReduce 版本 Join 程序，并能执行该程序和分析执行过程。

8.3　实验背景

8.3.1　概述

对于 RDBMS 中的 Join 操作大家一定非常熟悉，写 SQL 的时候要十分注意细节，稍有差池就会耗时很长造成很大的性能瓶颈，而在 Hadoop 中使用 MapReduce 框架进行 Join 的操作时同样耗时，但是由于 Hadoop 的分布式设计理念的特殊性，因此对于这种 Join 操作也具备了一定的特殊性。

8.3.2　原理

使用 MapReduce 实现 Join 操作有多种实现方式：

在 Reduce 端连接为最为常见的模式：

Map 端的主要工作：为来自不同表（文件）的 key/value 对打标签以区别不同来源的记录，然后用连接字段作为 key，其余部分和新加的标志作为 value，最后进行输出。

Reduce 端的主要工作：在 Reduce 端以连接字段作为 key 的分组已经完成，我们只需要在每一个分组当中将那些来源于不同文件的记录（在 map 阶段已经打标志）分开，最后进行笛卡尔就可以了。

在 Map 端进行连接：

使用场景：一张表十分小、一张表很大。

用法：在提交作业的时候首先将小表文件放到该作业的 DistributedCache 中，其次从 DistributeCache 中取出该小表进行 Join key / value 解释分割放到内存中（可以放大 Hash Map 等容器中）。最后扫描大表，看大表中的每条记录的 Join key /value 值是否能够在内存中找到相同 Join key 的记录，如果有则直接输出结果。

SemiJoin：

SemiJoin 就是所谓的半连接，其实仔细一看就是 Reduce Join 的一个变种，就是在 map 端过滤掉一些数据，在网络中只传输参与连接的数据，不参与连接的数据不必在网络中进行传输，从而减少了 shuffle 的网络传输量，使整体效率提高，其他思想和 Reduce Join 是一模一样的。也就是将小表中参与 Join 的 key 单独抽出来通过 DistributedCach 分发到相关节点，然后将其取出放到内存中（可以放到 HashSet 中），在 map 阶段扫描连接表，将 Join key 不在内存 HashSet 中的记录过滤掉，让那些参与 Join 的记录通过 shuffle 传输到 Reduce 端进行 Join 操作，其他和 Reduce Join 都一样。

8.4 实验步骤

8.4.1 准备阶段

这里，我们介绍最为常见的在 Reduce 端连接的代码编写流程。

首先准备数据，数据分为两个文件，分别为 A 表和 B 表数据：

A 表数据

```
201001 1003 abc
201002 1005 def
201003 1006 ghi
201004 1003 jkl
201005 1004 mno
201006 1005 pqr
```

B 表数据

```
1003 kaka
1004 da
1005 jue
1006 zhao
```

现在要通过程序得到 A 表第二个字段和 B 表第一个字段一致的数据的 Join 结果：

```
1003    201001 abc    kaka
1003    201004 jkl    kaka
1004    201005 mno da
1005    201002 def    jue
1005    201006 pqr jue
1006    201003 ghi zhao
```

程序分析执行过程如下：

在 map 阶段，把所有记录标记成<key, value>的形式，其中 key 是 1003/1004/1005/1006 的字段值，value 则根据来源不同取不同的形式：来源于表 A 的记录，value 的值为

“201001 abc”等值；来源于表 B 的记录，value 的值为“kaka”之类的值。

在 Reduce 阶段，先把每个 key 下的 value 列表拆分为分别来自表 A 和表 B 的两部分，分别放入两个向量中。然后遍历两个向量做笛卡尔积，形成一条条最终结果。

8.4.2　编写程序

完整代码：

```
package mr;

import java.io.DataInput;
import java.io.DataOutput;
import java.io.IOException;

import org.apache.hadoop.conf.Configuration;
import org.apache.hadoop.fs.Path;
import org.apache.hadoop.io.LongWritable;
import org.apache.hadoop.io.Text;
import org.apache.hadoop.io.WritableComparable;
import org.apache.hadoop.io.WritableComparator;
import org.apache.hadoop.mapreduce.Job;
import org.apache.hadoop.mapreduce.Mapper;
import org.apache.hadoop.mapreduce.Partitioner;
import org.apache.hadoop.mapreduce.Reducer;
import org.apache.hadoop.mapreduce.lib.input.FileInputFormat;
import org.apache.hadoop.mapreduce.lib.output.FileOutputFormat;
import org.apache.hadoop.mapreduce.lib.input.FileSplit;
import org.apache.hadoop.util.GenericOptionsParser;

public class MRJoin {
public static class MR_Join_Mapper extends Mapper<LongWritable, Text, TextPair, Text> {
        @Override
        protected void map(LongWritable key, Text value, Context context)
                                                    throws IOException, InterruptedException {
            // 获取输入文件的全路径和名称
            String pathName = ((FileSplit) context.getInputSplit()).getPath().toString();
            if (pathName.contains("data.txt")) {
                    String values[] = value.toString().split("\t");
                    if (values.length < 3) {
                        // data 数据格式不规范，字段小于 3，抛弃数据
                        return;
                    } else {
```

```
                // 数据格式规范，区分标识为 1
                TextPair tp = new TextPair(new Text(values[1]), new Text("1"));
                context.write(tp, new Text(values[0] + "\t" + values[2]));
            }
        }
        if (pathName.contains("info.txt")) {
            String values[] = value.toString().split("\t");
            if (values.length < 2) {
                // data 数据格式不规范，字段小于 2，抛弃数据
                return;
            } else {
                // 数据格式规范，区分标识为 0
                TextPair tp = new TextPair(new Text(values[0]), new Text("0"));
                context.write(tp, new Text(values[1]));
            }
        }
    }
}

public static class MR_Join_Partitioner extends Partitioner<TextPair, Text> {
    @Override
    public int getPartition(TextPair key, Text value, int numParititon) {
        return Math.abs(key.getFirst().hashCode() * 127) % numParititon;
    }
}

public static class MR_Join_Comparator extends WritableComparator {
    public MR_Join_Comparator() {
        super(TextPair.class, true);
    }

    public int compare(WritableComparable a, WritableComparable b) {
        TextPair t1 = (TextPair) a;
        TextPair t2 = (TextPair) b;
        return t1.getFirst().compareTo(t2.getFirst());
    }
}

public static class MR_Join_Reduce extends Reducer<TextPair, Text, Text, Text> {
    protected void Reduce(TextPair key, Iterable<Text> values, Context context)
            throws IOException, InterruptedException {
```

```
            Text pid = key.getFirst();
            String desc = values.iterator().next().toString();
            while (values.iterator().hasNext()) {
                context.write(pid, new Text(values.iterator().next().toString() + "\t" + desc));
            }
        }
    }


public static void main(String agrs[])
                        throws IOException, InterruptedException, ClassNotFoundException {
    Configuration conf = new Configuration();
    GenericOptionsParser parser = new GenericOptionsParser(conf, agrs);
    String[] otherArgs = parser.getRemainingArgs();
    if (agrs.length < 3) {
        System.err.println("Usage: MRJoin <in_path_one> <in_path_two> <output>");
        System.exit(2);
    }

    Job job = new Job(conf, "MRJoin");
    // 设置运行的 job
    job.setJarByClass(MRJoin.class);
    // 设置 Map 相关内容
    job.setMapperClass(MR_Join_Mapper.class);
    // 设置 Map 的输出
    job.setMapOutputKeyClass(TextPair.class);
    job.setMapOutputValueClass(Text.class);
    // 设置 partition
    job.setPartitionerClass(MR_Join_Partitioner.class);
    // 在分区之后按照指定的条件分组
    job.setGroupingComparatorClass(MR_Join_Comparator.class);
    // 设置 Reduce
    job.setReducerClass(MR_Join_Reduce.class);
    // 设置 Reduce 的输出
    job.setOutputKeyClass(Text.class);
    job.setOutputValueClass(Text.class);
    // 设置输入和输出的目录
    FileInputFormat.addInputPath(job, new Path(otherArgs[0]));
    FileInputFormat.addInputPath(job, new Path(otherArgs[1]));
    FileOutputFormat.setOutputPath(job, new Path(otherArgs[2]));
    // 执行，直到结束就退出
```

```
        System.exit(job.waitForCompletion(true) ? 0 : 1);
    }
}

class TextPair implements WritableComparable<TextPair> {
private Text first;
private Text second;

public TextPair() {
        set(new Text(), new Text());
    }

public TextPair(String first, String second) {
        set(new Text(first), new Text(second));
}

public TextPair(Text first, Text second) {
        set(first, second);
}

public void set(Text first, Text second) {
        this.first = first;
        this.second = second;
}

public Text getFirst() {
        return first;
}

  public Text getSecond() {
        return second;
  }

  public void write(DataOutput out) throws IOException {
        first.write(out);
        second.write(out);
  }

  public void readFields(DataInput in) throws IOException {
        first.readFields(in);
        second.readFields(in);
```

```
    }

    public int compareTo(TextPair tp) {
        int cmp = first.compareTo(tp.first);
        if (cmp != 0) {
            return cmp;
        }
        return second.compareTo(tp.second);
    }
}
```

8.4.3　打包并提交

使用 Eclipse 开发工具将该代码打包，假定打包后的文件名为 MRJoin.jar，主类 MRJoin 位于包 mr 下，则可使用如下命令向 Hadoop 集群提交本应用。

```
[root@client hadoop]# bin/hadoop jar MRJoin.jar mr.MRJoin /usr/MRJoin/in/data.txt /usr/MRJoin/in/info.txt /usr/MRJoin/out
```

其中“hadoop”为命令，“jar”为命令参数，后面紧跟打包。“/usr/MRJoin/in/data.txt”和“/usr/MRJoin/in/info.txt”为输入文件在 HDFS 中的位置，“/usr/MRJoin/out”为输出文件在 HDFS 中的位置。

8.5　实验结果

8.5.1　输入数据

输入数据如下：data.txt （数据统一放在/root/data 目录下）。

```
201001 1003 abc
201002 1005 def
201003 1006 ghi
201004 1003 jkl
201005 1004 mno
201006 1005 pqr
```

输入数据如下：info.txt （数据统一放在/root/data 目录下）。

```
1003 kaka
1004 da
1005 jue
1006 zhao
```

8.5.2　输出显示

在 client 上执行对 hdfs 上的文件/usr/MRJoin/out/part-r-00000 内容查看的操作。

```
[root@client hadoop]#   bin/hadoop   fs   -cat   /usr/MRJoin/out/p*
```

查看 MR Join 结果如图 8-1 所示。

```
[root@client hadoop]# bin/hadoop fs -cat /usr/MRjoin/out/p*
.16/11/01 13:28:56 WARN util.NativeCodeLoader: Unable to load
re applicable
1003    201004  jkl     kaka
1003    201001  abc     kaka
1004    201005  mno     da
1005    201006  pqr     jue
1005    201002  def     jue
1006    201003  ghi     zhao
[root@client hadoop]#
```

图 8-1　查看 MR Join 结果

实验九　MapReduce 实验：分布式缓存

9.1　实验目的

理解序列化与反序列化；熟悉 Configuration 类；学会使用 Configuration 类进行参数传递；学会在 Map 或 Reduce 阶段引用 Configuration 传来的参数；理解分布式缓存“加载小表、扫描大表”的处理思想。

9.2　实验要求

假定现有一个大为 100GB 的大表 big.txt 和一个大小为 1MB 的小表 small.txt，请基于 MapReduce 思想编程实现判断小表中单词在大表中出现次数。所谓的“扫描大表、加载小表”。

9.3　实验步骤

为解决上述问题，可开启 10 个 Map，这样每个 Map 只需处理总量的 1/10，将大大加快处理。而在单独 Map 内，直接用 HashSet 加载“1MB 小表”，对于存在硬盘（Map 处理时会将 HDFS 文件复制至本地）的 10GB 大文件，则逐条扫描，这就是所谓的“扫描大表、加载小表”，即分布式缓存，如图 9-1 所示。

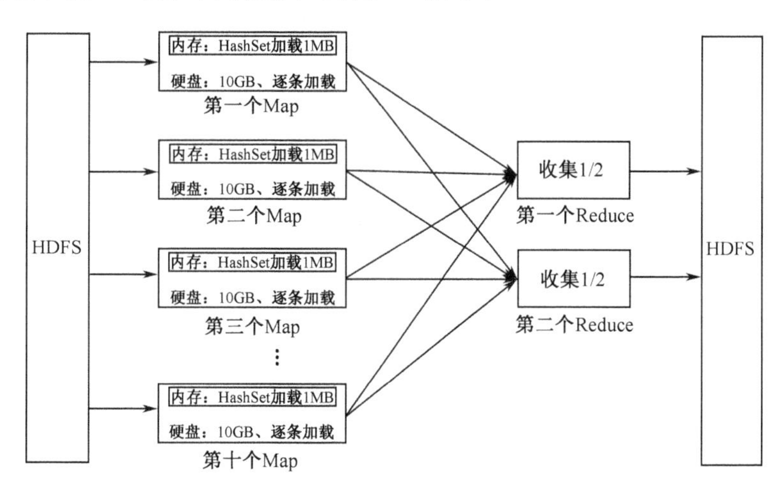

图 9-1　扫描大表、加载小表

由于实验中没有 100GB 这样的大表，甚至连 1MB 这样的小表都没有，因为本实验采用模拟方式，所以用少量数据代表大文件 big.txt，更少量数据代表 small.txt。整个实验步骤为“准备数据→上传数据→编写代码→执行代码→查看结果”这五大步骤。

9.3.1 准备数据

为降低操作难度，此处用少量数据代表大文件 big.txt，更少量数据代表小文件 small.txt，具体操作如下。

首先，登录 client 机，确认该机上存在“/root/data/9/big.txt”和“/root/data/9/small.txt”，如图 9-2 所示，显然 big.txt 内容为“aaa~zzz 和 000~999”，small.txt 为其中三项。

```
[root@client ~]# cat  /root/data/9/big.txt ←
aaa bbb ccc ddd eee fff ggg
hhh iii jjj kkk lll mmm nnn
000 111 222 333 444 555 666 777 888 999
ooo ppp qqq rrr sss ttt
uuu vvv www xxx yyy zzz
[root@client ~]# cat  /root/data/9/small.txt ←
eee sss 555
[root@client ~]#
```

图 9-2 确认本地文件 big.txt 和 small.txt

9.3.2 上传数据

首先，登录 client 机，查看 HDFS 里是否已存在目录“/user/root/mr/in”，若不存在，使用下述命令新建该目录。

```
[root@client ~]# /usr/cstor/hadoop/bin/hdfs  dfs  -mkdir  -p  /user/root/mr/in
```

其次，使用下述命令将 client 机本地文件“/root/data/9/big.txt”和“/root/data/9/small.txt”上传至 HDFS 的“/user/root/mr/in”目录：

```
[root@client ~]# /usr/cstor/hadoop/bin/hdfs  dfs  -put  /root/data/9/big.txt  /user/root/mr/in
[root@client ~]# /usr/cstor/hadoop/bin/hdfs  dfs  -put  /root/data/9/small.txt  /user/root/mr/in
```

最后，使用命令确认 HDFS 上文件与内容，如图 9-3 所示。

```
[root@client ~]# /usr/cstor/hadoop/bin/hdfs  dfs  -cat /user/root/mr/in/big.txt
aaa bbb ccc ddd eee fff ggg
hhh iii jjj kkk lll mmm nnn
000 111 222 333 444 555 666 777 888 999
ooo ppp qqq rrr sss ttt
uuu vvv www xxx yyy zzz
[root@client ~]# /usr/cstor/hadoop/bin/hdfs  dfs  -cat /user/root/mr/in/small.txt
eee sss 555
[root@client ~]#
```

图 9-3 确认 HDFS 文件 big.txt 和 small.txt

9.3.3 编写代码

首先，打开 Eclipse，依次点击“File→New→Other…→Map/Reduce Project”，在弹

出的“New MapReduce Project Wizard”对话框中，“Project name:”一栏填写项目名“BigSmallTable”，然后直接点击该对话框的“Finish”按钮，如图9-4所示。

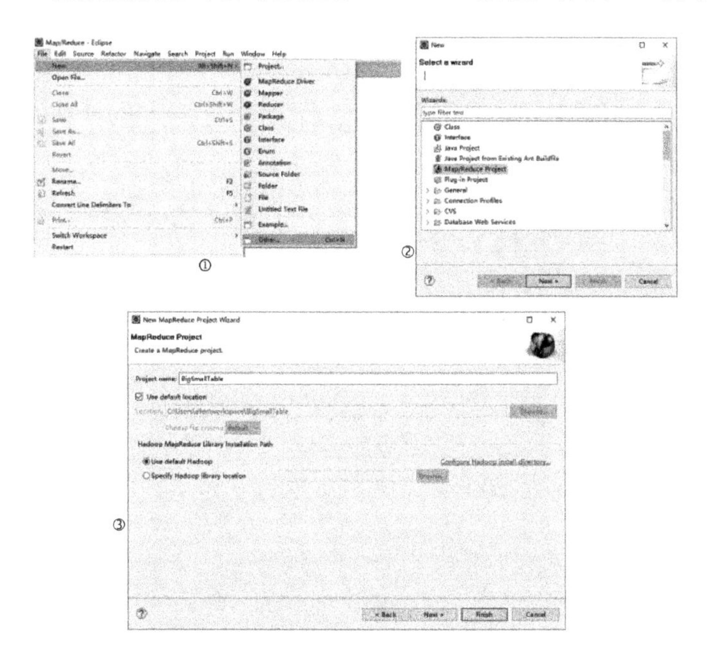

图9-4　确认文件big.txt和small.txt

其次，新建BigAndSmallTable类并指定包名（代码中为cn.cstor.mr），在BigAndSmallTable.java文件中，依次写入如下代码：

```
package cn.cstor.mr;

import java.io.IOException;
import java.util.HashSet;

import org.apache.hadoop.conf.Configuration;
import org.apache.hadoop.fs.FSDataInputStream;
import org.apache.hadoop.fs.FileSystem;
import org.apache.hadoop.fs.Path;
import org.apache.hadoop.io.IntWritable;
import org.apache.hadoop.io.Text;
import org.apache.hadoop.mapreduce.Job;
import org.apache.hadoop.mapreduce.Mapper;
import org.apache.hadoop.mapreduce.Reducer;
import org.apache.hadoop.mapreduce.lib.input.FileInputFormat;
import org.apache.hadoop.mapreduce.lib.output.FileOutputFormat;
import org.apache.hadoop.util.LineReader;

public class BigAndSmallTable {
```

```
public static class TokenizerMapper extends
Mapper<Object, Text, Text, IntWritable> {
private final static IntWritable one = new IntWritable(1);
private static HashSet<String> smallTable = null;

protected void setup(Context context) throws IOException,
InterruptedException {
smallTable = new HashSet<String>();
Path smallTablePath = new Path(context.getConfiguration().get(
"smallTableLocation"));
FileSystem hdfs = smallTablePath.getFileSystem(context
.getConfiguration());
FSDataInputStream hdfsReader = hdfs.open(smallTablePath);
Text line = new Text();
LineReader lineReader = new LineReader(hdfsReader);
while (lineReader.readLine(line) > 0) {
// you can do something here
String[] values = line.toString().split(" ");
for (int i = 0; i < values.length; i++) {
smallTable.add(values[i]);
System.out.println(values[i]);
}
}
lineReader.close();
hdfsReader.close();
System.out.println("setup ok *^_^* ");
}

public void map(Object key, Text value, Context context)
throws IOException, InterruptedException {
String[] values = value.toString().split(" ");
for (int i = 0; i < values.length; i++) {
if (smallTable.contains(values[i])) {
context.write(new Text(values[i]), one);
}
}
}
}

public static class IntSumReducer extends
Reducer<Text, IntWritable, Text, IntWritable> {
private IntWritable result = new IntWritable();

public void reduce(Text key, Iterable<IntWritable> values,
```

```
Context context) throws IOException, InterruptedException {
int sum = 0;
for (IntWritable val : values) {
sum += val.get();
}
result.set(sum);
context.write(key, result);
}
}

public static void main(String[] args) throws Exception {
Configuration conf = new Configuration();
conf.set("smallTableLocation", args[1]);
Job job = Job.getInstance(conf, "BigAndSmallTable");
job.setJarByClass(BigAndSmallTable.class);
job.setMapperClass(TokenizerMapper.class);
job.setReducerClass(IntSumReducer.class);
job.setMapOutputKeyClass(Text.class);
job.setMapOutputValueClass(IntWritable.class);
job.setOutputKeyClass(Text.class);
job.setOutputValueClass(IntWritable.class);
FileInputFormat.addInputPath(job, new Path(args[0]));
FileOutputFormat.setOutputPath(job, new Path(args[2]));
System.exit(job.waitForCompletion(true) ? 0 : 1);
}
}
```

图 9-5 为本项目开发过程示例，请读者对照该图，分析项目结构图中各模块。

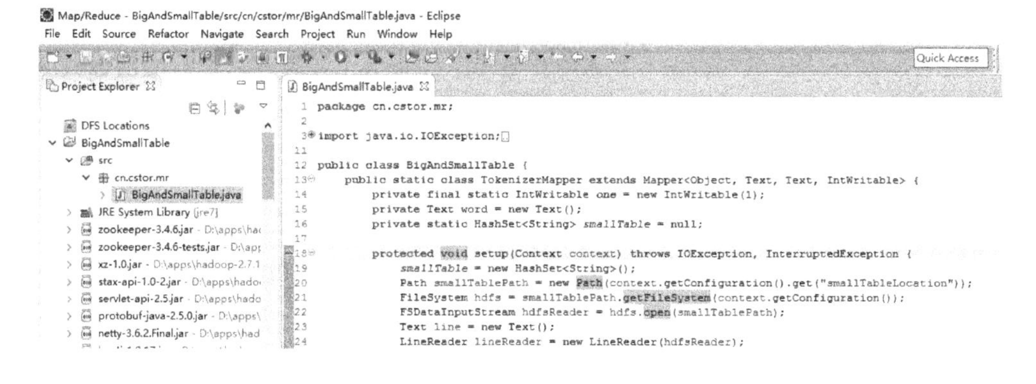

图 9-5　项目开发过程示例（一）

待代码编写结束，选中该项目，依次点击“Export→Java→JAR file”，在弹出对话框填写打包位置，接着 Finish 即可，如图 9-6 所示。笔者此处打包时包名及其位置为“C:\Users\allen\Desktop\BigSmallTable.jar”。

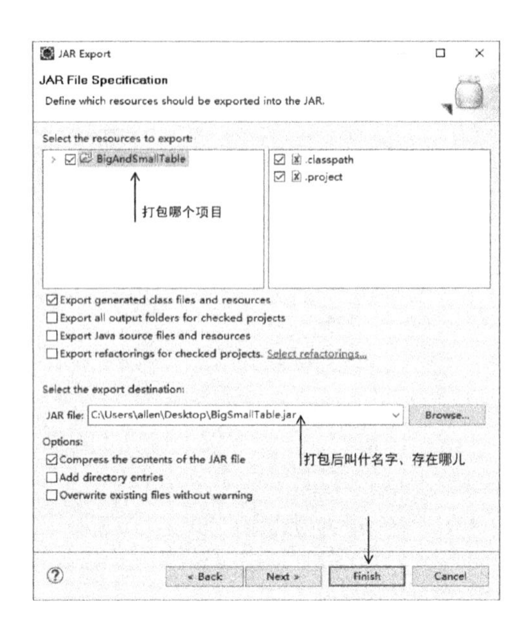

图 9-6　项目开发过程示例（二）

9.3.4　执行代码

首先，使用“Xmanager Enterprise 5”将“C:\Users\allen\ Desktop\BigSmallTable.jar”上传至 client 机。此处上传至“/root/BigSmallTable.jar”。

其次，登录 client 机上，使用下述命令提交 BigSmallTable.jar 任务。

```
[root@client ~]# /usr/cstor/hadoop/bin/hadoop  jar  /root/BigSmallTable.jar
  cn.cstor.mr.BigAndSmallTable  /user/root/mr/in/big.txt
  /user/root/mr/in/small.txt  /user/root/mr/bigAndSmallResult
```

9.3.5　查看结果

程序执行后，可使用下述命令查看执行结果，注意若再次执行，请更改结果目录：

```
[root@client ~]# /usr/cstor/hadoop/bin/hdfs  dfs  -cat  /user/root/mr/bigAndSmallResult/part-r-00000
```

9.4　实验结果

实验结果如图 9-7 所示，根据 big.txt，small.txt 文件内容和编程目的，可知实验结果准确无误。

```
[root@client ~]#
[root@client ~]# /usr/cstor/hadoop/bin/hdfs  dfs  -cat  /user/root/mr/bigAndSmallResult/part-r-00000
555     1
eee     1
sss     1
[root@client ~]#
```

图 9-7　查看实验结果

实验十　Hive 实验：部署 Hive

10.1　实验目的

1. 理解 Hive 存在的原因；
2. 理解 Hive 的工作原理；
3. 理解 Hive 的体系架构；
4. 学会如何进行内嵌模式部署；
5. 启动 Hive，然后将元数据存储在 HDFS 上。

10.2　实验要求

1. 完成 Hive 的内嵌模式部署；
2. 能够将 Hive 数据存储在 HDFS 上；
3. 待 Hive 环境搭建好后，能够启动并执行一般命令。

10.3　实验原理

Hive 是 Hadoop 大数据生态圈中的数据仓库，其提供以表格的方式来组织与管理 HDFS 上的数据、以类 SQL 的方式来操作表格里的数据，Hive 的设计目的是能够以类 SQL 的方式查询存放在 HDFS 上的大规模数据集，不必开发专门的 MapReduce 应用。

Hive 本质上相当于一个 MapReduce 和 HDFS 的翻译终端，用户提交 Hive 脚本后，Hive 运行时环境会将这些脚本翻译成 MapReduce 和 HDFS 操作并向集群提交这些操作。

当用户向 Hive 提交其编写的 HiveQL 后，首先，Hive 运行时环境会将这些脚本翻译成 MapReduce 和 HDFS 操作；其次，Hive 运行时环境使用 Hadoop 命令行接口向 Hadoop 集群提交这些 MapReduce 和 HDFS 操作；最后，Hadoop 集群逐步执行这些 MapReduce 和 HDFS 操作，整个过程可概括如下。

（1）用户编写 HiveQL 并向 Hive 运行时环境提交该 HiveQL。

（2）Hive 运行时环境将该 HiveQL 翻译成 MapReduce 和 HDFS 操作。

（3）Hive 运行时环境调用 Hadoop 命令行接口或程序接口，向 Hadoop 集群提交翻译后的 HiveQL。

（4）Hadoop 集群执行 HiveQL 翻译后的 MapReduce-APP 或 HDFS-APP。

由上述执行过程可知，Hive 的核心是其运行时环境，该环境能够将类 SQL 语句编译成 MapReduce。

Hive 构建在基于静态批处理的 Hadoop 之上，Hadoop 通常都有较高的延迟并且在作业提交和调度的时候需要大量的开销。因此，Hive 并不能够在大规模数据集上实现低延迟、快速的查询，例如，Hive 在几百 MB 的数据集上执行查询一般有分钟级的时间延迟。

因此，Hive 并不适合那些需要低延迟的应用，例如，联机事务处理（OLTP）。Hive 查询操作过程严格遵守 Hadoop MapReduce 的作业执行模型，Hive 将用户的 HiveQL 语句通过解释器转换为 MapReduce 作业提交到 Hadoop 集群上，Hadoop 监控作业执行过程，然后返回作业执行结果给用户。Hive 并非为联机事务处理而设计，Hive 并不提供实时的查询和基于行级的数据更新操作。Hive 的最佳使用场合是大数据集的批处理作业，例如，网络日志分析。

Hive 架构与基本组成如图 10-1 所示。

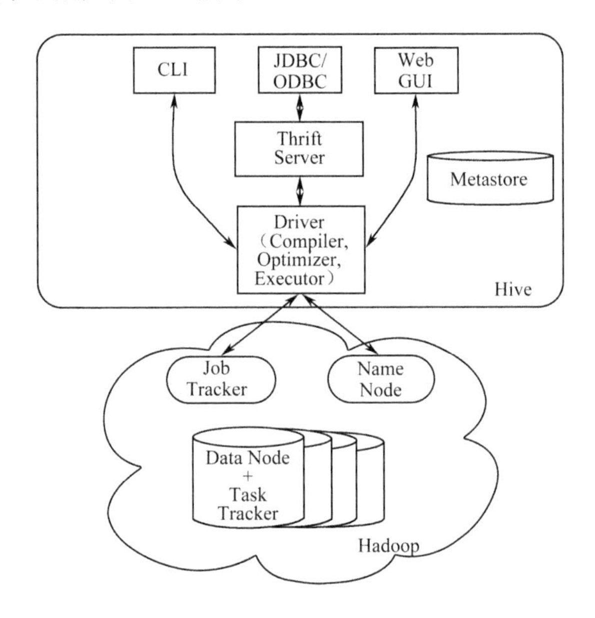

图 10-1　Hive 架构与基本组成

10.4　实验步骤

相对于其他组件，Hive 部署要复杂得多，按 metastore 存储位置的不同，其部署模式分为内嵌模式、本地模式和完全远程模式三种。当使用完全模式时，可以提供很多用户同时访问并操作 Hive，并且此模式还提供各类接口（BeeLine，CLI，甚至是 Pig），这里我们以内嵌模式为例。

由于使用内嵌模式时，其 Hive 会使用内置的 Derby 数据库来存储数据库，此时无须考虑数据库部署连接问题，整个部署过程可概括如下。

10.4.1　安装部署

在 client 机上操作：首先确定存在 Hive。

```
[root@client~]# ls /usr/cstor/
hive/
[root @client~]#
```

10.4.2　配置 HDFS

先为 Hive 配置 Hadoop 安装路径。待解压完成后，进入 Hive 的配置文件夹 conf 目录下，接着将 Hive 的环境变量模板文件复制成环境变量文件。

```
[root@client~]# cd /usr/cstor/hive/conf
[root@client conf]# cp hive-env.sh.template hive-env.sh
[root@client conf]# vim hive-env.sh
```

在配置文件中加入以下语句：

```
HADOOP_HOME=/usr/cstor/hadoop
```

然后在 HDFS 里新建 Hive 的存储目录。在 HDFS 中新建 /tmp 和 /usr/hive/warehouse 两个文件目录，并对同组用户增加写权限。

```
[root@client hadoop]# bin/hadoop fs -mkdir /tmp
[root@client hadoop]# bin/hadoop fs -mkdir -p /usr/hive/warehouse
[root@client hadoop]# bin/hadoop fs -chmod g+w /tmp
[root@client hadoop]# bin/hadoop fs -chmod g+w /usr/hive/warehouse
```

10.4.3　启动 Hive

在内嵌模式下，启动 Hive 指的是启动 Hive 运行时环境，用户可使用下述命令进入 Hive 运行时环境。

启动 Hive 命令行：

```
[root@client ~]# cd /usr/cstor/hive/
[root@client hive ]# bin/hive
```

10.5　实验结果

10.5.1　启动结果

使用“bin/Hive”命令进入 Hive 环境验证 Hive 是否启动成功。如图 10-2 所示。

```
[root@client Hive ]# bin/Hive
```

```
[root@client hive]# bin/hive

Logging initialized using configuration in jar:
hive>
```

图 10-2　验证 Hive 是否启动成功

10.5.2 Hive基本命令

使用“bin/Hive”命令进入 Hive 环境后，使用 “show tables” “show function”后，结果如图 10-3 和图 10-4 所示，则表示配置成功。

```
Hive> show tables ;
```

```
hive> show tables;
OK
Time taken: 1.148 seconds
hive>
```

图 10-3 show tabls 结果

```
Hive> show functions ;
```

```
weekofyear
when
windowingtablefunction
xpath
xpath_boolean
xpath_double
xpath_float
xpath_int
xpath_long
xpath_number
xpath_short
xpath_string
year
|
~
Time taken: 0.045 seconds, Fetched: 216 row(s)
hive>
```

图 10-4 show function 结果

实验十一　Hive 实验：新建 Hive 表

11.1　实验目的

1. 创建 Hive 的表；
2. 显示 Hive 中的所有表；
3. 显示 Hive 中表的列项；
4. 修改 Hive 中的表并能够删除 Hive 中的表。

11.2　实验要求

1. 完成 Hive 的 DDL 操作；
2. 在 Hive 中新建、显示、修改和删除表等功能。

11.3　实验原理

Hive 没有专门的数据存储格式，也没有为数据建立索引，用户可以非常自由地组织 Hive 中的表，只需要在创建表的时候告诉 Hive 数据中的列分隔符和行分隔符，Hive 就可以解析数据。

Hive 中所有的数据都存储在 HDFS 中，Hive 中包含以下数据模型：表（Table），外部表（External Table），分区（Partition），桶（Bucket）。

Hive 中 Table 和数据库中 Table 在概念上是类似的，每一个 Table 在 Hive 中都有一个相应的目录存储数据。例如，一个表 pvs，它在 HDFS 中的路径为：/wh/pvs，其中，wh 是在 hive-site.xml 中由${hive.metastore.warehouse.dir}指定的数据仓库的目录，所有的 Table 数据（不包括 External Table）都保存在这个目录中。

11.4　实验步骤

11.4.1　启动 Hive

启动 Hive 命令行：

```
[root@client ~]# cd /usr/cstor/hive/
[root@client hive ]# bin/hive
```

11.4.2 创建表

默认情况下，新建表的存储格式均为 Text 类型，字段间默认分隔符为键盘上的 Tab 键。

创建一个有两个字段的 pokes 表，其中第一列名为 foo，数据类型为 INT；第二列名为 bar，类型为 STRING：

```
hive> CREATE TABLE pokes (foo INT, bar STRING) ;
```

创建一个有两个实体列和一个（虚拟）分区字段的 invites 表：

```
hive> CREATE TABLE invites (foo INT, bar STRING) PARTITIONED BY (ds STRING) ;
注意：分区字段并不属于 invites，当向 invites 导入数据时，ds 字段会用来过滤导入的数据。
```

11.4.3 显示表

显示所有的表：

```
hive> SHOW TABLES ;
```

显示表（正则查询），同 MySQL 中操作一样，Hive 也支持正则查询，比如显示以.s 结尾的表：

```
hive> SHOW TABLES '.*s';
```

11.4.4 显示表列

显示表列：

```
hive> DESCRIBE invites;
```

11.4.5 更改表

修改表 events 名为 3koobecaf （自行创建任意类型 events 表）：

```
hive> ALTER TABLE events RENAME TO 3koobecaf;
```

将 pokes 表新增一列（列名为 new_col，类型为 INT）：

```
hive> ALTER TABLE pokes ADD COLUMNS (new_col INT);
```

将 invites 表新增一列（列名为 new_col2，类型为 INT），同时增加注释“a comment”：

```
hive> ALTER TABLE invites ADD COLUMNS (new_col2 INT COMMENT 'a comment');
```

替换 invites 表所有列名（数据不动）：

```
hive> ALTER TABLE invites REPLACE COLUMNS (foo INT, bar STRING, baz INT COMMENT 'baz replaces new_col2');
```

11.4.6 删除表（或列）

删除 invites 表 bar 和 baz 两列：

```
hive> ALTER TABLE invites REPLACE COLUMNS (foo INT COMMENT 'only keep the first column');
```

删除 pokes 表：

```
hive> DROP TABLE pokes;
```

11.5　实验结果

上述实验步骤的结果如图 11-1～图 11-11 所示。

创建一个有两个字段的 pokes 表，如图 11-1 所示。

```
hive> CREATE TABLE pokes (foo INT, bar STRING) ;
OK
Time taken: 1.262 seconds
hive>
```

图 11-1　创建 pokes 表

创建一个有两个实体列和一个（虚拟）分区字段的 invites 表，如图 11-2 所示。

```
hive> CREATE TABLE invites (foo INT, bar STRING) PARTITIONED BY (ds STRING) ;
OK
Time taken: 0.105 seconds
hive>
```

图 11-2　创建 invites 表

显示所有的表，如图 11-3 所示。

以.s 结尾的表，如图 11-4 所示。

```
hive> SHOW TABLES ;
OK
invites
pokes
Time taken: 0.198 seconds, Fetched: 2 row(s)
hive>
```

图 11-3 显示所有的表

```
hive> SHOW TABLES '.*s';
OK
invites
pokes
Time taken: 0.066 seconds, Fetched: 2 row(s)
hive>
```

图 11-4　以 s 结尾的表

显示表列，如图 11-5 所示。

```
hive> DESCRIBE invites;
OK
foo                     int
bar                     string
ds                      string

# Partition Information
# col_name              data_type               comment

ds                      string
Time taken: 0.29 seconds, Fetched: 8 row(s)
hive>
```

图 11-5　显示表列

修改表 events 名为 3koobecaf()自行创建任意类型 events 表，如图 11-6 所示。

```
hive> ALTER TABLE events RENAME TO 3koobecaf;
OK
Time taken: 0.255 seconds
hive>
```

图 11-6　表重命名

将 pokes 表新增一列（列名为 new_col，类型为 INT），如图 11-7 所示。

```
hive> ALTER TABLE pokes ADD COLUMNS (new_col INT);
OK
Time taken: 1.455 seconds
hive>
```

图 11-7　表 pokes 新增列

将 invites 表新增一列（列名为 new_col2，类型为 INT），如图 11-8 所示。

```
hive> ALTER TABLE invites ADD COLUMNS (new_col2 INT COMMENT 'a comment');
OK
Time taken: 0.142 seconds
hive>
```

图 11-8　表 invites 新增列

替换 invites 表所有列名，如图 11-9 所示。

```
hive> ALTER TABLE invites REPLACE COLUMNS (foo INT, bar STRING, baz INT COMMENT 'baz replaces new_col2');
OK
Time taken: 0.134 seconds
hive>
```

图 11-9　替换 invites 表的列名

删除 invites 表 bar 和 baz 两列，如图 11-10 所示。

```
hive> ALTER TABLE invites REPLACE COLUMNS (foo INT COMMENT 'only keep the first column');
OK
Time taken: 0.143 seconds
hive>
```

图 11-10　删除表列

删除 pokes 表，如图 11-11 所示。

```
hive> DROP TABLE pokes;
OK
Time taken: 1.17 seconds
hive>
```

图 11-11　删除表

实验十二 Hive 实验：Hive 分区

12.1 实验目的

掌握 Hive 分区的用法，加深对 Hive 分区概念的理解，了解 Hive 表在 HDFS 的存储目录结构。

12.2 实验要求

创建一个 Hive 分区表；根据数据年份创建 year=2014 和 year=2015 两个分区；将 2015 年的数据导入到 year=2015 的分区；在 Hive 界面用条件 year=2015 查询 2015 年的数据。

12.3 实验原理

分区（Partition）对应于数据库中的分区（Partition）列的密集索引，但是 Hive 中分区（Partition）的组织方式和数据库中的很不相同。在 Hive 中，表中的一个分区（Partition）对应于表下的一个目录，所有的分区（Partition）的数据都存储在对应的目录中。例如：pvs 表中包含 ds 和 ctry 两个分区（Partition），则对应于 ds = 20090801, ctry = US 的 HDFS 子目录为：/wh/pvs/ds=20090801/ctry=US；对应于 ds = 20090801, ctry = CA 的 HDFS 子目录为：/wh/pvs/ds=20090801/ctry=CA。

外部表（External Table）指向已经在 HDFS 中存在的数据，可以创建分区（Partition）。它和 Table 在元数据的组织上是相同的，而实际数据的存储则有较大的差异。

Table 的创建过程和数据加载过程（这两个过程可以在同一个语句中完成），在加载数据的过程中，实际数据会被移动到数据仓库目录中；之后对数据的访问将会直接在数据仓库目录中完成。删除表时，表中的数据和元数据将会被同时删除。

12.4 实验步骤

因为 Hive 依赖于 MapReduce，所以本实验之前先要启动 Hadoop 集群，然后再启动 Hive 进行实验，主要包括以下三个步骤。

12.4.1 启动 Hadoop 集群

在主节点进入 Hadoop 安装目录，启动 Hadoop 集群。

```
[root@master ~]# cd /usr/cstor/hadoop/sbin
[root@master sbin]#   ./start-all.sh
```

12.4.2 用命令进入 Hive 客户端

进入 Hive 安装目录，用命令进入 Hive 客户端。

```
[root@master ~]# cd   /usr/cstor/hive
[root@master hive]#   bin/hive
```

12.4.3 通过 HQL 语句进行实验

进入客户端后，查看 Hive 数据库，并选择 default 数据库：

```
hive> show databases;
 OK
 default
 Time taken: 1.152 seconds, Fetched: 1 row(s)
 hive> use default;
 OK
 Time taken: 0.036 seconds
在命令端创建 Hive 分区表：
hive> create table parthive (createdate string, value string) partitioned by (year string) row format delimited
fields terminated by '\t';
 OK
 Time taken: 0.298 seconds
```

查看新建的表：

```
hive> show tables;
 OK
 parthive
 Time taken: 1.127 seconds, Fetched: 1 row(s)
```

给 parthive 表创建两个分区：

```
hive> alter table parthive add partition(year='2014');
OK
Time taken: 0.195 seconds
hive> alter table parthive add partition(year='2015');
OK
Time taken: 0.121 seconds
```

查看 parthive 的表结构：

```
hive> describe parthive;
OK
createdate             string
value                    string
year                     string
# Partition Information
```

```
# col_name              data_type               comment
  year                  string
Time taken: 0.423 seconds, Fetched: 8 row(s)
```

向 year=2015 分区导入本地数据：

```
hive> load data local inpath '/root/data/12/parthive.txt' into table parthive partition(year='2015');
Loading data to table default.parthive partition (year=2015)
    Partition default.parthive{year=2015} stats: [numFiles=1, totalSize=110]
OK
Time taken: 1.071 seconds
```

根据条件查询 year=2015 的数据：

```
hive> select * from parthive t where t.year='2015';
```

根据条件统计 year=2015 的数据：

```
hive> select count(*) from parthive where year='2015';
```

12.5 实验结果

Hive 客户端查询结果如图 12-1 所示。

```
Time taken: 1.703 seconds
hive> select * from parthive t where t.year='2015';
OK
2015-01-01      aaa     2015
2015-02-01      bbb     2015
2015-03-01      ccc     2015
2015-04-01      ddd     2015
2015-05-01      eee     2015
2015-06-01      fff     2015
2015-07-01      ggg     2015
Time taken: 0.762 seconds, Fetched: 7 row(s)
```

图 12-1 客户端查询结果

Hive 客户端统计结果如图 12-2 所示。

```
hive> select count(*) from parthive where year='2015';
Query ID = root_20161129122609_3b522fb0-409b-49b8-b6e2-0f7f2f8087f3
Total jobs = 1
Launching Job 1 out of 1
Number of reduce tasks determined at compile time: 1
In order to change the average load for a reducer (in bytes):
  set hive.exec.reducers.bytes.per.reducer=<number>
In order to limit the maximum number of reducers:
  set hive.exec.reducers.max=<number>
In order to set a constant number of reducers:
  set mapreduce.job.reduces=<number>
Starting Job = job_1480180131746_0003, Tracking URL = http://master:8088/proxy/application_1480180131746_0003/
Kill Command = /usr/cstor/hadoop/bin/hadoop job  -kill job_1480180131746_0003
Hadoop job information for Stage-1: number of mappers: 1; number of reducers: 1
2016-11-29 12:26:15,852 Stage-1 map = 0%,  reduce = 0%
2016-11-29 12:26:22,179 Stage-1 map = 100%,  reduce = 0%, Cumulative CPU 1.68 sec
2016-11-29 12:26:29,533 Stage-1 map = 100%,  reduce = 100%, Cumulative CPU 4.12 sec
MapReduce Total cumulative CPU time: 4 seconds 120 msec
Ended Job = job_1480180131746_0003
MapReduce Jobs Launched:
Stage-Stage-1: Map: 1  Reduce: 1   Cumulative CPU: 4.12 sec   HDFS Read: 7082 HDFS Write: 2 SUCCESS
Total MapReduce CPU Time Spent: 4 seconds 120 msec
OK
7
Time taken: 20.996 seconds, Fetched: 1 row(s)
```

图 12-2 客户端统计结果

实验十三 Spark 实验：部署 Spark 集群

13.1 实验目的

能够理解 Spark 存在的原因，了解 Spark 的生态圈，理解 Spark 体系架构并理解 Spark 计算模型。学会部署 Spark 集群并启动 Spark 集群，能够配置 Spark 集群使用 HDFS。

13.2 实验要求

要求实验结束时，每位学生均已构建出以 Spark 集群：master 上部署主服务 Master；slave1、slave2、slave3 上部署从服务 Worker；client 上部署 Spark 客户端。待集群搭建好后，还需在 client 上进行下述操作：提交并运行 Spark 示例代码 WordCount，将 client 上某文件上传至 HDFS 里刚才新建的目录。

13.3 实验原理

13.3.1 Spark 简介

Spark 是一个高速的通用型集群计算框架，其内部内嵌了一个用于执行 DAG（有向无环图）的工作流引擎，能够将 DAG 类型的 Spark-App 拆分成 Task 序列并在底层框架上运行。在程序接口层，Spark 为当前主流语言都提供了编程接口，如用户可以使用 Scala、Java、Python、R 等高级语言直接编写 Spark-App。此外，在核心层之上，Spark 还提供了诸如 SQL、Mllib、GraphX、Streaming 等专用组件，这些组件内置了大量专用算法，充分利用这些组件，能够大大加快 Spark-App 开发进度。

一般称 Spark Core 为 Spark，Spark Core 处于存储层和高层组建层之间，定位为计算引擎，核心功能是并行化执行用户提交的 DAG 型 Spark-App。目前，Spark 生态圈主要包括 Spark Core 和基于 Spark Core 的独立组件（SQL、Streaming、Mllib 和 Graphx）。

13.3.2 Spark 适用场景

（1）Spark 是基于内存的迭代计算框架，适用于需要多次操作特定数据集的应用场合。

（2）由于 RDD 的特性，Spark 不适用那种异步细粒度更新状态的应用，例如 Web 服务的存储或者是增量的 Web 爬虫和索引。

（3）数据量不是特别大，但是要求实时统计分析需求。

13.4 实验步骤

13.4.1 配置 Spark 集群

配置 Spark 集群（独立模式）：

前提：1. 请自行配置各节点之间的免密登录，并在/etc/hosts 中写好 hostname 与 IP 的对应，这样方便配置文件的相互复制。2. 因为下面实验涉及 Spark 集群使用 HDFS，所以按照之前的实验预先部署好 HDFS。

在 master 机上操作：确定存在 Spark。

```
[root@master ~]# ls /usr/cstor
spark/
[root@master ~]#
```

在 master 机上操作：进入/usr/cstor 目录中。

```
[root@master ~]# cd /usr/cstor
[root@master cstor]#
```

进入配置文件目录/usr/cstor/spark/conf，先复制并修改 slave.templae 为 slave。

```
[root@master ~]# cd /usr/cstor/spark/conf
[root@master cstor]# cp  slaves.template slaves
```

然后，用 vim 命令编辑器编辑 slaves 文件。

```
[root@master cstor]# vim slaves
```

编辑 slaves 文件将下述内容添加到 slaves 文件中。

```
slave1
slave2
slave3
```

上述内容表示当前的 Spark 集群共有三台 slave 机，这三台机器的机器名称分别是 slave1~ slave3。

在 spark-conf.sh 中加入 JAVA_HOME。

```
[root@master cstor]# vim /usr/cstor/spark/sbin/spark-config.sh
```

加入以下内容：

```
export JAVA_HOME=/usr/local/jdk1.7.0_79
```

将配置好的 Spark 复制至 slaveX、client。（machines 在目录/root/data/2 下，如果不存在则自己新建一个）。

使用 for 循环语句完成多机复制。

```
[root@master ~]# cd /root/data/2
[root@master ~]# cat  machines
```

```
slave1
slave2
slave3
client
[root@master ~]# for  x  in  `cat  machines` ; do  echo  $x ; scp  -r  /usr/cstor/spark/ $x:/usr/cstor/; done;
```

在 master 机上操作：启动 Spark 集群。

```
[root@master local]# /usr/cstor/spark/sbin/start-all.sh
```

13.4.2 配置 HDFS

配置 Spark 集群使用 HDFS：

首先，关闭集群（在 master 上执行）。

```
[root@master ~]# /usr/cstor/spark/sbin/stop-all.sh
```

将 Spark 环境变量模板复制成环境变量文件。

```
[root@master ~]# cd /usr/cstor/spark/conf
[root@master conf]# cp spark-env.sh.template spark-env.sh
```

修改 Spark 环境变量配置文件 spark-env.sh。

```
[root@master conf]$ vim spark-env.sh
```

在 sprak-env.sh 配置文件中添加下列内容。

```
export HADOOP_CONF_DIR=/usr/cstor/hadoop/etc/hadoop
```

重新启动 spark。

```
[root@master local]# /usr/cstor/spark/sbin/start-all.sh
```

13.4.3 提交 Spark 任务

在 client 机上操作：使用 Shell 命令向 Spark 集群提交 Spark-App。

（1）上传 in.txt 文件到 HDFS（hdfs://master:8020/user/spark/in/）上。

in.txt 文件在/root/data/13/目录下。

请大家参照实验一自行完成。

（2）提交 WordCount 示例代码。

进入/usr/cstor/spark 目录，执行如下命令:

bin/spark-submit –master spark://master:7077 --class org.apache.spark.examples.JavaWordCount lib/spark-examples-1.6.0-hadoop2.6.0.jar hdfs://master:8020/user/spark/in/in.txt

```
[root@master local]# cd /usr/cstor/spark
[root@client spark]# bin/spark-submit --master spark://master:7077 \
> --class org.apache.spark.examples.JavaWordCount \
> lib/spark-examples-1.6.0-hadoop2.6.0.jar hdfs://master:8020/user/spark/in/in.txt
```

13.5 实验结果

13.5.1 进程查看

在 master 和 slave1-slave3 上分别执行 jps 命令查看对应进程。master 中进程为 Master，slave 机进程为 Worker，如图 13-1 所示。

```
[root@master spark-1.6.0]# jps
22122 Master
21210 ResourceManager
20960 SecondaryNameNode
20705 NameNode
24101 Jps
[root@master spark-1.6.0]#
```

```
[root@slave1 hadoop]# jps
9364 Worker
7820 DataNode
7959 NodeManager
10575 Jps
[root@slave1 hadoop]#
```

图 13-1 进程情况

13.5.2 验证 WebUI

在本地（需开启 OpenVPN）浏览器中输入 master 的 IP 和端口号 8080（如 10.1.89.5：8080），即可看到 Spark 的 WebUI。此页面包含了 Spark 集群主节点、从节点等各类统计信息，如图 13-2 所示。

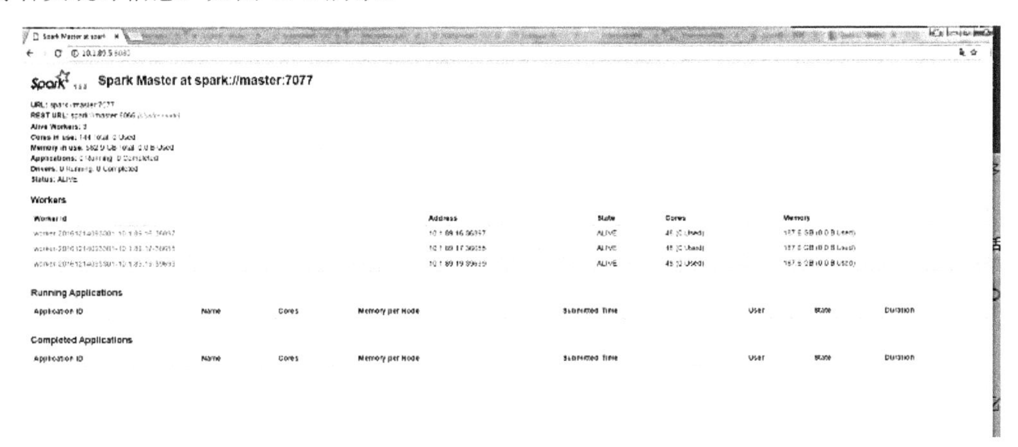

图 13-2 Spark-Web 页面

13.5.3 SparkWordcount 程序执行

输入：in.txt（数据放在/root/date/13 目录下）。

```
hello world
ni hao
hello my friend
ni are my sunshine
```

输出结果如图 13-3 所示。

```
16/11/04 13:36:06 INFO DAGScheduler: ResultStage 1 (collect at JavaWordCount.java:68) finished in 0.126 s
16/11/04 13:36:06 INFO DAGScheduler: Job 0 finished: collect at JavaWordCount.java:68, took 6.390418 s
are: 1
sunshine: 1
hello: 2
my: 2
friend: 1
ni: 2
hao: 1
world: 1
16/11/04 13:36:06 INFO SparkUI: Stopped Spark web UI at http://172.17.0.4:4040
16/11/04 13:36:06 INFO SparkDeploySchedulerBackend: Shutting down all executors
16/11/04 13:36:06 INFO SparkDeploySchedulerBackend: Asking each executor to shut down
16/11/04 13:36:06 INFO MapOutputTrackerMasterEndpoint: MapOutputTrackerMasterEndpoint stopped!
```

图 13-3　输出结果

WebUI 中 Application 的详细信息如图 13-4 所示。

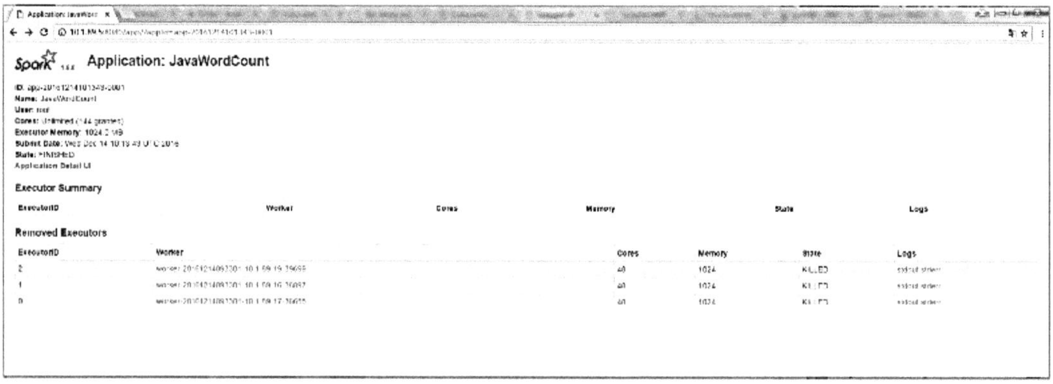

图 13-4　Web 中显示 Application 的详细信息

实验十四　Spark 实验：SparkWordCount

14.1　实验目的

熟悉 Scala 语言，基于 Spark 思想，编写 SparkWordCount 程序。

14.2　实验要求

熟悉 Scala 语言，理解 Spark 编程思想，并会编写 Spark 版本的 WordCount，然后能够在 Spark-shell 中执行代码和分析执行过程。

14.3　实验原理

Scala 是一门以 Java 虚拟机（JVM）为目标运行环境并将面向对象（OO）和函数式编程语言（FP）的最佳特性结合在一起的编程语言。

它既有动态语言那样的灵活简洁，同时又保留了静态类型检查带来的安全保障和执行效率，加上其强大的抽象能力，既能处理脚本化的临时任务，又能处理高并发场景下的分布式互联网大数据应用，可谓能缩能伸。

Scala 运行在 JVM 之上，因此它可以访问任何 Java 类库并且与 Java 框架进行互操作。其与 Java 的集成度很高，可以直接使用 Java 社区大量成熟的技术框架和方案。由于它直接编译成 Java 字节码，因此我们可以充分利用 JVM 这个高性能的运行平台。

14.3.1　Scala 是兼容的

Scala 被设计成无缝地与 Java 实施互操作。不需要你从 Java 平台后退两步然后跳到 Java 语言前面去。它允许你在现存代码中加点儿东西——在你已有的东西上建设，Scala 程序会被编译为 JVM 的字节码。它们的执行期性能通常与 Java 程序一致。Scala 代码可以调用 Java 方法，访问 Java 字段，继承自 Java 类和实现 Java 接口。这些都不需要特别的语法，显式接口描述，或粘接代码。实际上，几乎所有 Scala 代码都极度依赖于 Java 库，而程序员无须意识到这点 。

交互式操作的另一个方面是 Scala 极度重用了 Java 类型 。Scala 的 Int 类型代表了 Java 的原始整数类型 int，Float 代表了 float，Boolean 代表 boolean 等。Scala 的数组被映射到 Java 数组。Scala 同样重用了许多标准 Java 库类型。例如，Scala 里的字串文本“abc”是 java.lang.String，而抛出的异常必须是 java.lang.Throwable 的子类。

Scala 不仅重用了 Java 的类型，还把它们“打扮”得更漂亮 。例如，Scala 的字串支持类似于 toInt 和 toFloat 的方法，可以把字串转换成整数或者浮点数。因此你可以写 str.toInt 替代 Integer.parseInt（str）。如何在不打破互操作性的基础上做到这点呢？Java 的 String 类当然不会有 toInt 方法。实际上，Scala 有一个解决这种高级库设计和互操作性不相和谐的通用方案。Scala 可以让你定义隐式转换：implicit conversion，这常常用在类型失配，或者选用不存在的方法时。在上面的例子里，当在字串中寻找 toInt 方法时，Scala 编译器会发现 String 类里没有这种方法，但它会发现一个把 Java 的 String 转换为 Scala 的 RichString 类的一个实例的隐式转换，里面定义了这么个方法。于是在执行 toInt 操作之前，转换被隐式应用。

Scala 代码同样可以由 Java 代码调用。有时这种情况更加微妙，因为 Scala 是一种比 Java 更丰富的语言，有些 Scala 更先进的特性在它们能映射到 Java 前需要先被编码一下。

14.3.2 Scala 是简洁的

Scala 程序一般都很短。Scala 程序员曾报告说与 Java 比起来代码行数可以减少到 1/10，这有可能是个极限的例子。较保守的估计大概标准的 Scala 程序应该有 Java 写的同样的程序一半行数左右。更少的行数不仅意味着更少的打字工作，同样意味着更少的话在阅读和理解程序上的努力及更少的出错可能。许多因素在减少代码行上起了作用 。Scala 的语法避免了一些束缚 Java 程序的固定写法 。例如，Scala 里的分号是可选的，且通常不写。Scala 语法里还有很多其他的地方省略了东西。比方说，比较一下你在 Java 和 Scala 里是如何写类及构造函数的。

在 Java 里，带有构造函数的类经常看上去是这个样子：

```
// 在 Java 里
class MyClass {
    private int index;
    private String name;
    public MyClass(int index, String name) {
        this.index = index;
        this.name = name;
    }
}
```

在 Scala 里，你会写成这样：

```
class MyClass(index: Int, name: String)
```

根据这段代码，Scala 编译器将制造有两个私有成员变量的类，一个名为 index 的 Int 类型和一个叫做 name 的 String 类型，还有一个用这些变量作为参数获得初始值的构造函数。这个构造函数还将用作为参数传入的值初始化这两个成员变量。Scala 类写起来更快，读起来更容易，最重要的是，比 Java 类更不容易犯错。

使 Scala 简洁易懂的另一个因素是它的类型推断。重复的类型信息可以被忽略，因此程序变得更有条理和易读，但或许减少代码最关键原因的是已经存在于你的库里而不需要写的代码。Scala 给了你许多工具来定义强有力的库让你抓住并提炼出通用的行为。例如，库类的不同方面可以被分成若干特质，而这些又可以被灵活地混合在一起。或者，库方法可以用操作符参数化，从而让你有效地定义那些你自己控制的构造。这些构造组合在一起，就能够让库的定义既是高层级的又能灵活运用。

14.3.3 Scala 是高级的

程序员总是在和复杂性纠缠。为了高产出地编程，你必须明白你工作的代码。过度复杂的代码成了很多软件工程崩溃的原因。遗憾的是，重要的软件往往有复杂的需求，这种复杂性不可避免，必须（由不受控）转为受控。

Scala 可以通过提升你设计和使用的接口的抽象级别来帮助你管理复杂性。例如，假设你有一个 String 变量 name，你想弄清楚是否 String 包含一个大写字符。

在 Java 里，你或许这么写：

```
// 在 Java 里
boolean nameHasUpperCase = false;
for (int i = 0; i < name.length(); ++i) {
    if (Character.isUpperCase(name.charAt(i))) {
        nameHasUpperCase = true;
        break;
    }
}
```

在 Scala 里，你可以写成：

```
val nameHasUpperCase = name.exists(_.isUpperCase)
```

Java 代码把字符串看作循环中逐字符步进的低层级实体。Scala 代码把同样的字串当作能用论断。显然，Scala 代码更短并且对训练有素的眼睛来说，比 Java 代码更容易懂。因此 Scala 代码在通盘复杂度预算上能极度地变轻，它也更少给你机会犯错。

论断，_.isUpperCase，是一个 Scala 里面函数式文本的例子。它描述了带一个字符参量（用下画线字符代表）的函数，并测试其是否为大写字母。原则上，这种控制的抽象在 Java 中也是可能的，为此需要定义一个包含抽象功能的方法的接口。例如，如果你想支持对字串的查询，就应引入一个只有一个方法 hasProperty 的接口 CharacterProperty：

```
// 在 Java 里
interface CharacterProperty {
    boolean hasProperty(char ch);
}
```

然后你可以在 Java 里用这个接口格式一个方法 exists：它带一个字串和一个 CharacterProperty 参数，结果返回真如果字串中有某个字符符合属性。然后你可以这样调用 exists：

```
// 在 Java 里
exists(name, new CharacterProperty {
    boolean hasProperty(char ch) {
```

```
            return Character.isUpperCase(ch);
        }
    });
```

然而，所有这些真的感觉很复杂，复杂到实际上多数 Java 程序员都不会惹这个麻烦。他们会宁愿写个循环并漠视他们代码里复杂性的累加。此外，Scala 里的函数式文本真的很轻量，于是就频繁被使用。随着对 Scala 的逐步了解，你会发现有越来越多定义和使用你自己的控制抽象的机会。你将发现这能帮助避免代码重复并因此保持程序简短和清晰。

14.3.4 Scala 是静态类型的

静态类型系统认定变量和表达式与它们持有和计算的值的种类有关 。Scala 坚持作为一种具有非常先进的静态类型系统的语言。从 Java 那样的内嵌类型系统起步，能够让你使用泛型：generics 参数化类型，用交集：intersection 联合类型，用抽象类型：abstract type 隐藏类型的细节。这些为建造和组织你自己的类型打下了坚实的基础，从而能够设计出既安全又能灵活使用的接口。静态类型系统的经典优越性将更被赏识，其中最重要的包括程序抽象的可检验属性、安全的重构，以及更好的文档。

可检验属性。静态类型系统可以保证消除某些运行时的错误。例如，可以保证这样的属性：布尔型不会与整数型相加；私有变量不会从类的外部被访问；函数带了正确个数的参数；只有字串可以被加到字串集之中。不过当前的静态类型系统还不能查到其他类型的错误。比方说，通常查不到无法终结的函数，数组越界，或除零错误。同样也查不到你的程序不符合式样书（假设有这么一份式样书）。静态类型系统因此被认为不很有用而被忽视。舆论认为既然静态类型系统只能发现简单错误，而通过单元测试能提供发现更多的错误场景，又为何自寻烦恼呢？我们认为这种论调不对头。尽管静态类型系统确实不能替代单元测试，但是却能减少单元测试的数量。同样，单元测试也不能替代静态类型。总而言之，如 Edsger Dijkstra 所说，测试只能证明存在错误。因此，静态类型能给的保证或许很简单，但它们是无论多少测试都不能给的真正的保证。

安全的重构。静态类型系统提供了让你具有高度信心改动代码基础的安全网。试想一个对方法加入额外参数的重构实例。在静态类型语言中，你可以完成修改，重新编译你的系统并容易修改所有引起类型错误的代码行。一旦你完成了这些，你确信已经发现了所有需要修改的地方。对其他的简单重构，如改变方法名或把方法从一个类移到另一个，这种确信都有效。所有例子中静态类型检查会提供足够的确认，表明新系统和旧系统可以一样的工作 。

文档。静态类型是被编译器检查过正确性的程序文档。不像普通的注释，类型标注永远都不会过期（至少如果包含它的源文件近期刚刚通过编译就不会）。更进一步说，编译器和集成开发环境可以利用类型标注提供更好的上下文帮助。举例来说，集成开发环境可以通过判定选中表达式的静态类型，找到类型的所有成员，并全部显示出来。

虽然静态类型对程序文档来说通常很有用，当它们弄乱程序时，也会显得很突兀。标准意义上来说，有用的文档是读者不可能很容易地从程序中自己想出来的。在如下的方法定义中：

```
def f(x: String) = ...
```

知道 f 的变量应该是 String 是有用的。另外，以下例子中两个标注至少有一个是多余的：

```
val x: HashMap[Int, String] = new HashMap[Int, String]()
```

很明显，x 是以 Int 为键，String 为值的 HashMap 这句话无须重复两遍。

Scala 里的类型推断可以走的很远 。实际上，就算用户代码丝毫没有显式类型也不稀奇。因此，Scala 编程经常看上去有点像是动态类型脚本语言写出来的程序。尤其显著表现在作为粘接已写完的库控件的客户应用代码上。而对库控件来说不是这么回事，因为它们常常用到相当精妙的类型去使其适于灵活使用的模式。综上，构成可重用控件接口的成员类型符号应该是显式给出的，因为它们构成了控件和使用者间契约的重要部分。

14.4　实验步骤

在 Spark-shell 中编写 WordCount 代码和运行，上传 in.txt 文件到 HDFS 上。

请大家参照实验一自行完成。

启动 Spark-shell。

```
[root@client spark]# cd /usr/cstor/spark
[root@client spark]# bin/spark-shell --master spark://master:7077
```

写入 wordcount 的 scala 代码并运行。

```
scala> val file=sc.textFile("hdfs://master:8020/user/spark/in/in.txt")
scala> val count=file.flatMap(line => line.split(" ")).map(word => (word,1)).reduceByKey(_+_)
scala> count.collect()
```

14.5　实验结果

输入(in.txt) （数据统一放在/root/data/14 目录下）。

```
hello world
ni hao
hello my friend
ni are my sunshine
```

结束后运行结果如图 14-1 所示。

```
scala> val file=sc.textFile("hdfs://master:8020/usr/spark/in/in.txt")
file: org.apache.spark.rdd.RDD[String] = MapPartitionsRDD[1] at textFile at <console>:27

scala> val count=file.flatMap(line => line.split(" ")).map(word => (word,1)).reduceByKey(_+_)
count: org.apache.spark.rdd.RDD[(String, Int)] = ShuffledRDD[4] at reduceByKey at <console>:29

scala> count.collect()
res0: Array[(String, Int)] = Array((are,1), (hello,2), (my,2), (friend,1), (hao,1), (world,1), (sunshine,1), (ni,2))

scala> 
```

图 14-1　运行结果

实验十五　Spark 实验：RDD 综合实验

15.1　实验目的

1. 通过 Spark-shell 的操作理解 RDD 操作；
2. 能通过 RDD 操作的执行理解 RDD 的原理；
3. 对 Scala 有一定的认识。

15.2　实验要求

在实验结束时能完成 max，first，distinct，foreach 等 api 的操作。

15.3　实验原理

RDD（Resilient Distributed Datasets，弹性分布式数据集）是一个分区的只读记录的集合。RDD 只能通过在稳定的存储器或其他 RDD 数据上的确定性操作来创建。我们把这些操作称为变换以区别其他类型的操作。例如 map、filter 和 join。

RDD 在任何时候都不需要被“物化”（进行实际的变换并最终写入稳定的存储器上）。实际上，一个 RDD 有足够的信息描述着其如何从其他稳定的存储器上的数据生成。它有一个强大的特性：从本质上说，若 RDD 失效且不能重建，程序将不能引用该 RDD。而用户可以控制 RDD 的其他两个方面：持久化和分区。用户可以选择重用哪个 RDD，并为其制定存储策略，如内存存储。也可以让 RDD 中的数据根据记录的 key 分布到集群的多个机器。这对位置优化来说是有用的，比如可用来保证两个要 jion 的数据集都使用了相同的哈希分区方式。

通过 Spark 编程接口，编程人员可以对稳定存储上的数据进行变换操作（如 map 和 filter），得到一个或多个 RDD，然后可以调用这些 RDD 的 actions（动作）类的操作。这类操作的目是返回一个值或是将数据导入到存储系统中。动作类的操作如 count（返回数据集的元素数），collect（返回元素本身的集合）和 save（输出数据集到存储系统）。Spark 直到 RDD 第一次调用一个动作时才真正计算 RDD。

还可以调用 RDD 的 persist（持久化）方法来表明该 RDD 在后续操作中还会用到。默认情况下，Spark 会将调用过 persist 的 RDD 存在内存中。但若内存不足，也可以将其写入到硬盘上。通过指定 persist 函数中的参数，用户也可以请求其他持久化策略（如 Tachyon）并通过标记来进行 persist，比如仅存储到硬盘上或是在各机器之间复制一份。

最后，用户可以在每个 RDD 上设定一个持久化的优先级来指定内存中的哪些数据应该被优先写入到磁盘。缓存有个缓存管理器，Spark 里被称作 blockmanager。注意，这里还有一个误区，很多人认为调用了 cache 或者 persist 的那一刻就是在缓存了，这是不对的，真正的缓存执行指挥在 action 被触发。

总结：RDD 是分布式只读且已分区集合对象。这些集合是弹性的，如果数据集一部分丢失，则可以对它们进行重建。具有自动容错、位置感知调度和可伸缩性，而容错性是最难实现的，大多数分布式数据集的容错性有两种方式：数据检查点和记录数据的更新。对于大规模数据分析系统，数据检查点操作成本高，主要原因是大规模数据在服务器之间的传输带来的各方面的问题，相比记录数据的更新，RDD 也只支持粗粒度的转换，也就是记录如何从其他 RDD 转换而来（lineage），以便恢复丢失的分区。

简而言之，特性如下：

1. 数据结构不可变；
2. 支持跨集群的分布式数据操作；
3. 可对数据记录按 key 进行分区；
4. 提供了粗粒度的转换操作；
5. 数据存储在内存中，保证了低延迟性。

15.4 实验步骤

依据前面实验启动 Hadoop 和 Spark 集群。

利用 xmanager 中的 xshell 登录到 client 机器上。

进入到 client 机器的 Spark 的安装目录，执行命令：bin/spark-shell。

稍等一段时间，待 Spark-shell 启动之后，屏幕上出现 scala 的命令提示符之后开始进行 Spark 命令的键入。

注意：实验需要以本地模式启动：bin/spark-shell，如果以集群模式启动，有可能无法查看输出结果。

15.4.1 distinct 去除 RDD 内的重复数据

```
scala > var a = sc.parallelize(List("Gnu","Cat","Rat","Dog","Gnu","Rat"),2);
scala > a.distinct.collect
```

执行输出结果如图 15-1 所示。

```
scala> var a=sc.parallelize(List("Gnu","Cat","Rat","Dog","Gnu","Rat"),2);
a: org.apache.spark.rdd.RDD[String] = ParallelCollectionRDD[0] at parallelize at <console>:27

scala> a.distinct.collect
res0: Array[String] = Array(Dog, Cat, Gnu, Rat)

scala> 
```

图 15-1 执行结果（一）

15.4.2 foreach 遍历 RDD 内的数据

```
scala >var b = sc.parallelize(List(”cat”,”dog”,”tiger”,”lion”,”gnu”,”crocodile”,”ant”,”whale”,”dolphin”,”spider”),3)
scala > b.foreach(x=>println(x+”s are yummy”))
```

执行输出结果如图 15-2 所示。

```
scala> var b =sc.parallelize(List("cat","dog","tiger","lion","gnu","crocodile","ant","whale","dolphin","spider"),3);
b: org.apache.spark.rdd.RDD[String] = ParallelCollectionRDD[4] at parallelize at <console>:27

scala> b.foreach(x=>println(x+"s are yummy"));
[Stage 2:>                                                  (0 + 0) / 3]lions are yummy
gnus are yummy
crocodiles are yummy
ants are yummy
whales are yummy
dolphins are yummy
cats are yummy
spiders are yummy
dogs are yummy
tigers are yummy

scala>
```

图 15-2 执行结果（二）

15.4.3 first 取得 RDD 中的第一个数据

```
scala > var c=sc.parallelize(List(”dog”,”Cat”,”Rat”,”Dog”),2)
scala > c.first
```

执行输出结果如图 15-3 所示。

```
scala> var c=sc.parallelize(List("dog","cat","rat","dog"),2)
c: org.apache.spark.rdd.RDD[String] = ParallelCollectionRDD[5] at parallelize at <console>:27

scala> c.first
res2: String = dog

scala>
```

图 15-3 执行结果（三）

15.4.4 max 取得 RDD 中的最大的数据

```
scala > var d=sc.parallelize(10 to 30)
scala > d.max
```

执行输出结果如图 15-4 所示。

```
scala> var d=sc.parallelize(10 to 30)
d: org.apache.spark.rdd.RDD[Int] = ParallelCollectionRDD[6] at parallelize at <console>:27

scala> d.max
res3: Int = 30

scala>
```

图 15-4 执行结果（四）

```
scala > var e = sc.parallelize(List((10, "dog"),(20, "cat"),(30, "tiger"),(18, "lion")))
scala > e.max
```

执行输出结果如图 15-5 所示。

```
scala> var e=sc.parallelize(List((10,"dog"),(20,"cat"),(30,"tiger"),(18,"lion")))
e: org.apache.spark.rdd.RDD[(Int, String)] = ParallelCollectionRDD[7] at parallelize at <console>:27

scala> e.max
res4: (Int, String) = (30,tiger)

scala>
```

图 15-5　执行结果（五）

15.4.5　intersection 返回两个 RDD 重叠的数据

```
scala > var f = sc.parallelize(1 to 20)
scala > var g = sc.parallelize(10 to 30)
scala > var h = f.intersection(g)
scala > h.collect
```

执行输出结果如图 15-6 所示。

```
scala> var f=sc.parallelize(1 to 20)
f: org.apache.spark.rdd.RDD[Int] = ParallelCollectionRDD[24] at parallelize at <console>:27

scala> var g = sc.parallelize(10 to 30)
g: org.apache.spark.rdd.RDD[Int] = ParallelCollectionRDD[25] at parallelize at <console>:27

scala> var h = f.intersection(g)
h: org.apache.spark.rdd.RDD[Int] = MapPartitionsRDD[31] at intersection at <console>:31

scala> h.collect
res9: Array[Int] = Array(10, 11, 12, 13, 14, 15, 16, 17, 18, 19, 20)

scala>
```

图 15-6　执行结果（六）

15.5　实验结果

本次实验结果均已在实验步骤中给出。

实验十六　Spark 实验：Spark 综例

16.1　实验目的

1. 理解 Spark 编程思想；
2. 在 Spark Shell 中编写 Scala 程序；
3. 在 Spark Shell 中运行 Scala 程序。

16.2　实验要求

实验结束后，能够编写 Scala 代码解决以下问题，并能够自行分析执行过程。

有三个 RDD，要求统计 rawRDDA 中“aa”“bb”两个单词出现的次数；要求对去重后的 rawRDDA 再去掉 rawRDDB 中的内容；最后将上述两个结果合并成同一个文件然后存入 HDFS 中。

16.3　实验原理

16.3.1　Scala

Scala 是一门多范式的编程语言，一种类似 Java 的编程语言，设计初衷是实现可伸缩的语言，并集成面向对象编程和函数式编程的各种特性。

Scala 有几项关键特性表明了它的面向对象的本质。例如，Scala 中的每个值都是一个对象，包括基本数据类型（布尔值、数字等）在内，连函数也是对象。另外，类可以被子类化，而且 Scala 还提供了基于 mixin 的组合（mixin-based composition）。

与只支持单继承的语言相比，Scala 具有更广泛意义上的类重用。Scala 允许定义新类的时候重用“一个类中新增的成员定义（相较于其父类的差异之处）”。Scala 称之为 mixin 类组合。

Scala 还包含了若干函数式语言的关键概念，包括高阶函数（Higher-Order Function）、局部套用（Currying）、嵌套函数（Nested Function）、序列解读（Sequence Comprehensions）等。

Scala 是静态类型的，这就允许它提供泛型类、内部类，甚至多态方法（Polymorphic Method）。另外值得一提的是，Scala 被特意设计成能够与 Java 和.NET

互操作。

Scala 可以与 Java 互操作。它用 scalac 这个编译器把源文件编译成 Java 的 class 文件。你可以从 Scala 中调用所有的 Java 类库，也同样可以从 Java 应用程序中调用 Scala 的代码。

这让 Scala 得以使用为 Java1.4、5.0 或者 6.0 编写的巨量的 Java 类库和框架，Scala 会经常性地针对这几个版本的 Java 进行测试。Scala 可能也可以在更早版本的 Java 上运行，但没有经过正式的测试。Scala 以 BSD 许可发布，并且数年前就已经被认为相当稳定了。

Scala 旨在提供一种编程语言，能够统一和一般化分别来自面向对象和函数式两种不同风格的关键概念。借着这个目标与设计，Scala 得以提供一些出众的特性：

（1）面向对象风格

（2）函数式风格

（3）更高层的并发模型

Scala 把 Erlang 风格的基于 actor 的并发带进了 JVM。开发者可以利用 Scala 的 actor 模型在 JVM 上设计具伸缩性的并发应用程序，它会自动获得多核心处理器带来的优势，而不必依照复杂的 Java 线程模型来编写程序。

（4）轻量级的函数语法

高阶；

嵌套；

局部套用（Currying）；

匿名。

（5）与 XML 集成

可在 Scala 程序中直接书写 XML；

可将 XML 转换成 Scala 类。

（6）与 Java 无缝地互操作

总而言之，Scala 是一种函数式面向对象语言，它融汇了许多前所未有的特性，而同时又运行于 JVM 之上。

16.3.2 Spark-shell

该命令用于以交互式方式编写并执行 Spark App，且书写语法为 Scala。

下面的示例命令用于进入交互式执行器，进入执行器后，即可使用 Scala 语句以交互式方式编写并执行 Spark-App。

```
[root@client spark]# bin/spark-shell --master spark://master:7077
```

在该示例中，写明“--master spark://master:7077”的目的是使 Spark-shell 进入集群模式，若不写明，则 Spark-shell 会默认进入单机模式。

由于 Spark 使用 Scala 开发，而 Scala 实际上在 JVM 中执行，因此，我们搭建好 Spark 环境后，无须另外安装 Scala 组件。

16.4 实验步骤

16.4.1 启动 Spark–shell

登录 client 服务器，在集群模式下启动 Spark-shell，结果如图 16-1 所示。

```
[root@client ~]# cd /usr/cstor/spark/
[root@client spark]# bin/spark-shell --master spark://master:7077
```

```
client
[root@client spark]# bin/spark-shell --master spark://master:7077
16/12/14 09:23:02 WARN util.NativeCodeLoader: Unable to load native-hadoop library for your platform... using builtin-java classes where applicable
16/12/14 09:23:02 INFO spark.SecurityManager: Changing view acls to: root
16/12/14 09:23:02 INFO spark.SecurityManager: Changing modify acls to: root
16/12/14 09:23:02 INFO spark.SecurityManager: SecurityManager: authentication disabled; ui acls disabled; users with view permissions: Set(root); users with modify permissions: Set(root)
16/12/14 09:23:03 INFO spark.HttpServer: Starting HTTP Server
16/12/14 09:23:03 INFO server.Server: jetty-8.y.z-SNAPSHOT
16/12/14 09:23:03 INFO server.AbstractConnector: Started SocketConnector@0.0.0.0:36252
16/12/14 09:23:03 INFO util.Utils: Successfully started service 'HTTP class server' on port 36252.
Welcome to
      Spark   version 1.6.0

Using Scala version 2.10.5 (Java HotSpot(TM) 64-Bit Server VM, Java 1.7.0_79)
Type in expressions to have them evaluated.
Type :help for more information.
16/12/14 09:23:06 INFO spark.SparkContext: Running Spark version 1.6.0
16/12/14 09:23:06 INFO spark.SecurityManager: Changing view acls to: root
16/12/14 09:23:06 INFO spark.SecurityManager: Changing modify acls to: root
16/12/14 09:23:06 INFO spark.SecurityManager: SecurityManager: authentication disabled; ui acls disabled; users with view permissions: Set(root); users with modify permissions: Set(root)
16/12/14 09:23:06 INFO util.Utils: Successfully started service 'sparkDriver' on port 33613.
16/12/14 09:23:07 INFO slf4j.Slf4jLogger: Slf4jLogger started
16/12/14 09:23:07 INFO Remoting: Starting remoting
16/12/14 09:23:07 INFO Remoting: Remoting started; listening on addresses :[akka.tcp://sparkDriverActorSystem@10.1.1.20:45029]
16/12/14 09:23:07 INFO util.Utils: Successfully started service 'sparkDriverActorSystem' on port 45029.
16/12/14 09:23:07 INFO spark.SparkEnv: Registering MapOutputTracker
16/12/14 09:23:07 INFO spark.SparkEnv: Registering BlockManagerMaster
16/12/14 09:23:07 INFO storage.DiskBlockManager: Created local directory at /tmp/blockmgr-3ac24b1c-3a30-413b-8b29-40602bc5e40a
16/12/14 09:23:07 INFO storage.MemoryStore: MemoryStore started with capacity 511.5 MB
16/12/14 09:23:07 INFO spark.SparkEnv: Registering OutputCommitCoordinator
16/12/14 09:23:07 INFO server.Server: jetty-8.y.z-SNAPSHOT
16/12/14 09:23:07 INFO server.AbstractConnector: Started SelectChannelConnector@0.0.0.0:4040
16/12/14 09:23:07 INFO util.Utils: Successfully started service 'SparkUI' on port 4040.
16/12/14 09:23:07 INFO ui.SparkUI: Started SparkUI at http://10.1.1.20:4040
16/12/14 09:23:07 INFO client.AppClient$ClientEndpoint: Connecting to master spark://master:7077...
16/12/14 09:23:08 INFO cluster.SparkDeploySchedulerBackend: Connected to Spark cluster with app ID app-20161214092327-0000
16/12/14 09:23:08 INFO util.Utils: Successfully started service 'org.apache.spark.network.netty.NettyBlockTransferService' on port 41084.
16/12/14 09:23:08 INFO netty.NettyBlockTransferService: Server created on 41084
16/12/14 09:23:08 INFO storage.BlockManagerMaster: Trying to register BlockManager
16/12/14 09:23:08 INFO storage.BlockManagerMasterEndpoint: Registering block manager 10.1.1.20:41084 with 511.5 MB RAM, BlockManagerId(driver, 10.1.1.20, 41084)
16/12/14 09:23:08 INFO storage.BlockManagerMaster: Registered BlockManager
16/12/14 09:23:08 INFO client.AppClient$ClientEndpoint: Executor added: app-20161214092327-0000/0 on worker-20161214091427-10.1.89.26-34949 (10.1.89.26:34949) with 48 cores
16/12/14 09:23:08 INFO cluster.SparkDeploySchedulerBackend: Granted executor ID app-20161214092327-0000/0 on hostPort 10.1.89.26:34949 with 48 cores, 1024.0 MB RAM
16/12/14 09:23:08 INFO client.AppClient$ClientEndpoint: Executor added: app-20161214092327-0000/1 on worker-20161214091427-10.1.89.27-34424 (10.1.89.27:34424) with 48 cores
16/12/14 09:23:08 INFO cluster.SparkDeploySchedulerBackend: Granted executor ID app-20161214092327-0000/1 on hostPort 10.1.89.27:34424 with 48 cores, 1024.0 MB RAM
16/12/14 09:23:08 INFO client.AppClient$ClientEndpoint: Executor added: app-20161214092327-0000/2 on worker-20161214091427-10.1.89.22-36246 (10.1.89.22:36246) with 48 cores
16/12/14 09:23:08 INFO cluster.SparkDeploySchedulerBackend: Granted executor ID app-20161214092327-0000/2 on hostPort 10.1.89.22:36246 with 48 cores, 1024.0 MB RAM
16/12/14 09:23:08 INFO client.AppClient$ClientEndpoint: Executor updated: app-20161214092327-0000/0 is now RUNNING
16/12/14 09:23:08 INFO client.AppClient$ClientEndpoint: Executor updated: app-20161214092327-0000/2 is now RUNNING
16/12/14 09:23:08 INFO client.AppClient$ClientEndpoint: Executor updated: app-20161214092327-0000/1 is now RUNNING
16/12/14 09:23:08 INFO cluster.SparkDeploySchedulerBackend: SchedulerBackend is ready for scheduling beginning after reached minRegisteredResourcesRatio: 0.0
```

```
client
16/12/14 09:23:13 INFO storage.BlockManagerMasterEndpoint: Registering block manager 10.1.89.0:42866 with 511.5 MB RAM, BlockManagerId(0, 10.1.89.0, 42866)
16/12/14 09:23:14 INFO metastore.MetaStoreDirectSql: Using direct SQL, underlying DB is DERBY
16/12/14 09:23:14 INFO metastore.ObjectStore: Initialized ObjectStore
16/12/14 09:23:14 WARN metastore.ObjectStore: Version information not found in metastore. hive.metastore.schema.verification is not enabled so recording the schema version 1.2.0
16/12/14 09:23:14 WARN metastore.ObjectStore: Failed to get database default, returning NoSuchObjectException
16/12/14 09:23:14 INFO metastore.HiveMetaStore: Added admin role in metastore
16/12/14 09:23:14 INFO metastore.HiveMetaStore: Added public role in metastore
16/12/14 09:23:14 INFO metastore.HiveMetaStore: No user is added in admin role, since config is empty
16/12/14 09:23:14 INFO metastore.HiveMetaStore: 0: get_all_databases
16/12/14 09:23:14 INFO HiveMetaStore.audit: ugi=root	ip=unknown-ip-addr	cmd=get_all_databases
16/12/14 09:23:14 INFO metastore.HiveMetaStore: 0: get_functions: db=default pat=*
16/12/14 09:23:14 INFO HiveMetaStore.audit: ugi=root	ip=unknown-ip-addr	cmd=get_functions: db=default pat=*
16/12/14 09:23:14 INFO DataNucleus.Datastore: The class "org.apache.hadoop.hive.metastore.model.MResourceUri" is tagged as "embedded-only" so does not have its own datastore table.
16/12/14 09:23:15 INFO session.SessionState: Created HDFS directory: /tmp/hive/root
16/12/14 09:23:15 INFO session.SessionState: Created local directory: /tmp/root
16/12/14 09:23:15 INFO session.SessionState: Created local directory: /tmp/091970bb-4dd4-4ae2-8c5e-44cba1fde26a_resources
16/12/14 09:23:15 INFO session.SessionState: Created HDFS directory: /tmp/hive/root/091970bb-4dd4-4ae2-8c5e-44cba1fde26a
16/12/14 09:23:15 INFO session.SessionState: Created local directory: /tmp/root/091970bb-4dd4-4ae2-8c5e-44cba1fde26a
16/12/14 09:23:15 INFO session.SessionState: Created HDFS directory: /tmp/hive/root/091970bb-4dd4-4ae2-8c5e-44cba1fde26a/_tmp_space.db
16/12/14 09:23:15 INFO hive.HiveContext: default warehouse location is /user/hive/warehouse
16/12/14 09:23:15 INFO hive.HiveContext: Initializing HiveMetastoreConnection version 1.2.1 using Spark classes.
16/12/14 09:23:15 INFO client.ClientWrapper: Inspected Hadoop version: 2.6.0
16/12/14 09:23:15 INFO client.ClientWrapper: Loaded org.apache.hadoop.hive.shims.Hadoop23Shims for Hadoop version 2.6.0
16/12/14 09:23:16 INFO metastore.HiveMetaStore: 0: Opening raw store with implemenation class:org.apache.hadoop.hive.metastore.ObjectStore
16/12/14 09:23:16 INFO metastore.ObjectStore: ObjectStore, initialize called
16/12/14 09:23:16 INFO DataNucleus.Persistence: Property datanucleus.cache.level2 unknown - will be ignored
16/12/14 09:23:16 INFO DataNucleus.Persistence: Property hive.metastore.integral.jdo.pushdown unknown - will be ignored
16/12/14 09:23:16 WARN DataNucleus.Connection: BoneCP specified but not present in CLASSPATH (or one of dependencies)
16/12/14 09:23:16 WARN DataNucleus.Connection: BoneCP specified but not present in CLASSPATH (or one of dependencies)
16/12/14 09:23:18 INFO metastore.ObjectStore: Setting MetaStore object pin classes with hive.metastore.cache.pinobjtypes="Table,StorageDescriptor,SerDeInfo,Partition,Database,Type,FieldSchema,Order"
16/12/14 09:23:18 INFO DataNucleus.Datastore: The class "org.apache.hadoop.hive.metastore.model.MFieldSchema" is tagged as "embedded-only" so does not have its own datastore table.
16/12/14 09:23:18 INFO DataNucleus.Datastore: The class "org.apache.hadoop.hive.metastore.model.MOrder" is tagged as "embedded-only" so does not have its own datastore table.
16/12/14 09:23:19 INFO DataNucleus.Datastore: The class "org.apache.hadoop.hive.metastore.model.MFieldSchema" is tagged as "embedded-only" so does not have its own datastore table.
16/12/14 09:23:19 INFO DataNucleus.Datastore: The class "org.apache.hadoop.hive.metastore.model.MOrder" is tagged as "embedded-only" so does not have its own datastore table.
16/12/14 09:23:20 INFO metastore.MetaStoreDirectSql: Using direct SQL, underlying DB is DERBY
16/12/14 09:23:20 INFO metastore.ObjectStore: Initialized ObjectStore
16/12/14 09:23:20 WARN metastore.ObjectStore: Version information not found in metastore. hive.metastore.schema.verification is not enabled so recording the schema version 1.2.0
16/12/14 09:23:20 WARN metastore.ObjectStore: Failed to get database default, returning NoSuchObjectException
16/12/14 09:23:20 INFO metastore.HiveMetaStore: Added admin role in metastore
16/12/14 09:23:20 INFO metastore.HiveMetaStore: Added public role in metastore
16/12/14 09:23:20 INFO metastore.HiveMetaStore: No user is added in admin role, since config is empty
16/12/14 09:23:20 INFO metastore.HiveMetaStore: 0: get_all_databases
16/12/14 09:23:20 INFO HiveMetaStore.audit: ugi=root	ip=unknown-ip-addr	cmd=get_all_databases
16/12/14 09:23:20 INFO metastore.HiveMetaStore: 0: get_functions: db=default pat=*
16/12/14 09:23:20 INFO HiveMetaStore.audit: ugi=root	ip=unknown-ip-addr	cmd=get_functions: db=default pat=*
16/12/14 09:23:20 INFO DataNucleus.Datastore: The class "org.apache.hadoop.hive.metastore.model.MResourceUri" is tagged as "embedded-only" so does not have its own datastore table.
16/12/14 09:23:21 INFO session.SessionState: Created local directory: /tmp/21e34236-cf90-427d-a06d-73954e546840_resources
16/12/14 09:23:21 INFO session.SessionState: Created HDFS directory: /tmp/hive/root/21e34236-cf90-427d-a06d-73954e546840
16/12/14 09:23:21 INFO session.SessionState: Created local directory: /tmp/root/21e34236-cf90-427d-a06d-73954e546840
16/12/14 09:23:21 INFO session.SessionState: Created HDFS directory: /tmp/hive/root/21e34236-cf90-427d-a06d-73954e546840/_tmp_space.db
16/12/14 09:23:21 INFO repl.SparkILoop: Created sql context (with Hive support)..
SQL context available as sqlContext.

scala>
```

图 16-1　启动日志

16.4.2 编写并执行 Scala 代码

在 Spark-shell 执行器中编写如下 Scala 代码并运行，结果如图 16-2 所示。

```
scala> val rawRDDA = sc.parallelize(List("!!bb##cc", "%%ccbb%%", "cc&&++aa"), 3)
scala> val rawRDDB = sc.parallelize(List(("xx", 99), ("yy", 88), ("xx", 99), ("zz", 99)), 2)
scala> val rawRDDC = sc.parallelize(List(("yy",88)), 1)
scala> import org.apache.spark.HashPartitioner
scala> var tempResultRDDA = rawRDDA.flatMap(line=>line.split("")
                    ).filter(allWord=>{allWord.contains("aa") || allWord.contains("bb")}
                    ).map(word=>(word, 1)
                    ).partitionBy(new HashPartitioner(2)
                    ).groupByKey(
                    ).map((P:(String, Iterable[Int]))=>(P._1, P._2.sum))
scala> var tempResultRDDBC = rawRDDB.distinct.subtract(rawRDDC)
scala> var resultRDDABC = tempResultRDDA.union(tempResultRDDBC)
scala> resultRDDABC.saveAsTextFile("hdfs://master:8020/user/spark/resultRDDABC")
```

```
scala> val rawRDDA = sc.parallelize(List("!!bb##cc", "%%ccbb%%", "cc&&++aa"), 3)
rawRDDA: org.apache.spark.rdd.RDD[String] = ParallelCollectionRDD[0] at parallelize at <console>:27

scala> val rawRDDB = sc.parallelize(List(("xx", 99), ("yy", 88), ("xx", 99), ("zz", 99)), 2)
rawRDDB: org.apache.spark.rdd.RDD[(String, Int)] = ParallelCollectionRDD[1] at parallelize at <console>:27

scala> val rawRDDC = sc.parallelize(List(("yy",88)), 1)
rawRDDC: org.apache.spark.rdd.RDD[(String, Int)] = ParallelCollectionRDD[2] at parallelize at <console>:27

scala> import org.apache.spark.HashPartitioner
import org.apache.spark.HashPartitioner

scala> var tempResultRDDA = rawRDDA.flatMap(line=>line.split("")
     |                    ).filter(allWord=>{allWord.contains("aa") || allWord.contains("bb")}
     |                    ).map(word=>(word, 1)
     |                    ).partitionBy(new HashPartitioner(2)
     |                    ).groupByKey(
     |                    ).map((P:(String, Iterable[Int]))=>(P._1, P._2.sum))
tempResultRDDA: org.apache.spark.rdd.RDD[(String, Int)] = MapPartitionsRDD[8] at map at <console>:35

scala> var tempResultRDDBC = rawRDDB.distinct.subtract(rawRDDC)
tempResultRDDBC: org.apache.spark.rdd.RDD[(String, Int)] = MapPartitionsRDD[15] at subtract at <console>:32

scala> var resultRDDABC = tempResultRDDA.union(tempResultRDDBC)
resultRDDABC: org.apache.spark.rdd.RDD[(String, Int)] = UnionRDD[16] at union at <console>:38

scala> resultRDDABC.saveAsTextFile("hdfs://master:8020/user/spark/resultRDDABC")
16/12/14 09:58:22 INFO Configuration.deprecation: mapred.tip.id is deprecated. Instead, use mapreduce.task.id
16/12/14 09:58:22 INFO Configuration.deprecation: mapred.task.id is deprecated. Instead, use mapreduce.task.attempt.id
16/12/14 09:58:22 INFO Configuration.deprecation: mapred.task.is.map is deprecated. Instead, use mapreduce.task.ismap
16/12/14 09:58:22 INFO Configuration.deprecation: mapred.task.partition is deprecated. Instead, use mapreduce.task.partition
16/12/14 09:58:22 INFO Configuration.deprecation: mapred.job.id is deprecated. Instead, use mapreduce.job.id
16/12/14 09:58:22 INFO spark.SparkContext: Starting job: saveAsTextFile at <console>:41
```

图 16-2 代码执行截图

16.4.3 退出 Spark-shell

在执行器中执行下列命令，退出 Spark-shell，结果如图 16-3 所示。

```
scala> exit
```

```
scala> exit
warning: there were 1 deprecation warning(s); re-run with -deprecation for details
16/12/14 09:59:35 INFO spark.SparkContext: Invoking stop() from shutdown hook
16/12/14 09:59:35 INFO handler.ContextHandler: stopped o.s.j.s.ServletContextHandler{/static/sql,null}
16/12/14 09:59:35 INFO handler.ContextHandler: stopped o.s.j.s.ServletContextHandler{/SQL/execution/json,null}
16/12/14 09:59:35 INFO handler.ContextHandler: stopped o.s.j.s.ServletContextHandler{/SQL/execution,null}
16/12/14 09:59:35 INFO handler.ContextHandler: stopped o.s.j.s.ServletContextHandler{/SQL/json,null}
16/12/14 09:59:35 INFO handler.ContextHandler: stopped o.s.j.s.ServletContextHandler{/SQL,null}
16/12/14 09:59:35 INFO handler.ContextHandler: stopped o.s.j.s.ServletContextHandler{/metrics/json,null}
16/12/14 09:59:35 INFO handler.ContextHandler: stopped o.s.j.s.ServletContextHandler{/stages/stage/kill,null}
16/12/14 09:59:35 INFO handler.ContextHandler: stopped o.s.j.s.ServletContextHandler{/api,null}
16/12/14 09:59:35 INFO handler.ContextHandler: stopped o.s.j.s.ServletContextHandler{/,null}
16/12/14 09:59:35 INFO handler.ContextHandler: stopped o.s.j.s.ServletContextHandler{/static,null}
16/12/14 09:59:35 INFO handler.ContextHandler: stopped o.s.j.s.ServletContextHandler{/executors/threadDump/json,null}
16/12/14 09:59:35 INFO handler.ContextHandler: stopped o.s.j.s.ServletContextHandler{/executors/threadDump,null}
16/12/14 09:59:35 INFO handler.ContextHandler: stopped o.s.j.s.ServletContextHandler{/executors/json,null}
16/12/14 09:59:35 INFO handler.ContextHandler: stopped o.s.j.s.ServletContextHandler{/executors,null}
16/12/14 09:59:35 INFO handler.ContextHandler: stopped o.s.j.s.ServletContextHandler{/environment/json,null}
16/12/14 09:59:35 INFO handler.ContextHandler: stopped o.s.j.s.ServletContextHandler{/environment,null}
16/12/14 09:59:35 INFO handler.ContextHandler: stopped o.s.j.s.ServletContextHandler{/storage/rdd/json,null}
16/12/14 09:59:35 INFO handler.ContextHandler: stopped o.s.j.s.ServletContextHandler{/storage/rdd,null}
16/12/14 09:59:35 INFO handler.ContextHandler: stopped o.s.j.s.ServletContextHandler{/storage/json,null}
16/12/14 09:59:35 INFO handler.ContextHandler: stopped o.s.j.s.ServletContextHandler{/storage,null}
16/12/14 09:59:35 INFO handler.ContextHandler: stopped o.s.j.s.ServletContextHandler{/stages/pool/json,null}
16/12/14 09:59:35 INFO handler.ContextHandler: stopped o.s.j.s.ServletContextHandler{/stages/pool,null}
16/12/14 09:59:35 INFO handler.ContextHandler: stopped o.s.j.s.ServletContextHandler{/stages/stage/json,null}
16/12/14 09:59:35 INFO handler.ContextHandler: stopped o.s.j.s.ServletContextHandler{/stages/stage,null}
16/12/14 09:59:35 INFO handler.ContextHandler: stopped o.s.j.s.ServletContextHandler{/stages/json,null}
16/12/14 09:59:35 INFO handler.ContextHandler: stopped o.s.j.s.ServletContextHandler{/stages,null}
16/12/14 09:59:35 INFO handler.ContextHandler: stopped o.s.j.s.ServletContextHandler{/jobs/job/json,null}
16/12/14 09:59:35 INFO handler.ContextHandler: stopped o.s.j.s.ServletContextHandler{/jobs/job,null}
16/12/14 09:59:35 INFO handler.ContextHandler: stopped o.s.j.s.ServletContextHandler{/jobs/json,null}
16/12/14 09:59:35 INFO handler.ContextHandler: stopped o.s.j.s.ServletContextHandler{/jobs,null}
16/12/14 09:59:35 WARN thread.QueuedThreadPool: 4 threads could not be stopped
16/12/14 09:59:35 INFO ui.SparkUI: Stopped Spark web UI at http://10.1.89.21:4040
16/12/14 09:59:35 INFO cluster.SparkDeploySchedulerBackend: Shutting down all executors
16/12/14 09:59:35 INFO cluster.SparkDeploySchedulerBackend: Asking each executor to shut down
16/12/14 09:59:35 INFO spark.MapOutputTrackerMasterEndpoint: MapOutputTrackerMasterEndpoint stopped!
16/12/14 09:59:35 INFO storage.MemoryStore: MemoryStore cleared
16/12/14 09:59:35 INFO storage.BlockManager: BlockManager stopped
16/12/14 09:59:35 INFO storage.BlockManagerMaster: BlockManagerMaster stopped
16/12/14 09:59:35 INFO scheduler.OutputCommitCoordinator$OutputCommitCoordinatorEndpoint: OutputCommitCoordinator stopped!
16/12/14 09:59:35 INFO spark.SparkContext: Successfully stopped SparkContext
16/12/14 09:59:35 INFO util.ShutdownHookManager: Shutdown hook called
16/12/14 09:59:35 INFO util.ShutdownHookManager: Deleting directory /tmp/spark-1ceb55df-4981-4839-acb3-a3ebe789bedf
16/12/14 09:59:35 INFO util.ShutdownHookManager: Deleting directory /tmp/spark-f45bd2bf-1386-460d-9617-0895df21751c
16/12/14 09:59:35 INFO remote.RemoteActorRefProvider$RemotingTerminator: Shutting down remote daemon.
16/12/14 09:59:35 INFO remote.RemoteActorRefProvider$RemotingTerminator: Remote daemon shut down; proceeding with flushing remote transports.
16/12/14 09:59:35 INFO util.ShutdownHookManager: Deleting directory /tmp/spark-8b0d5c14-23a5-4c2d-9838-3836823d02c8
16/12/14 09:59:35 INFO remote.RemoteActorRefProvider$RemotingTerminator: Remoting shut down.
```

图 16-3　退出 Spark-shell

16.4.4　查看执行结果

执行 Hadoop 命令查看运行结果，结果如图 16-4 所示。

```
[root@client hadoop]# cd /usr/cstor/hadoop
[root@client hadoop]# bin/hadoop fs -cat /user/spark/resultRDDABC/p*
```

```
master ×
[root@master spark]# hadoop fs -cat /user/spark/resultRDDABC/p*
16/12/14 10:25:07 WARN util.NativeCodeLoader: Unable to load native-hadoop libra
ry for your platform... using builtin-java classes where applicable
(zz,99)
(xx,99)
[root@master spark]#
```

图 16-4　执行结果

实验十七　Spark 实验：Spark SQL

17.1　实验目的

1. 了解 Spark SQL 所能实现的功能；
2. 使用 Spark SQL 执行一些 SQL 语句。

17.2　实验要求

1. 在实验结束之后建立数据库，建立数据表的数据结构；
2. 建立数据表后在 Spark SQL 中执行 SQL 语句进行查询；
3. 向 Spark SQL 中导入数据。

17.3　实验原理

Spark SQL 用于以交互式方式编写并执行 Spark SQL，且书写语法为类 SQL，同 Spark-shell 一样，启动时写明“--master spark://master:7077”则进入集群模式，否则默认进入单机模式。由于默认安装的 Spark 已经包含了 Spark SQL，故无须安装其他组件，直接执行即可。

Spark SQL 使运行 SQL 和 HiveQL 查询十分简单。Spark SQL 能够轻易地定位相应的表和元数据。Spark SQL 为 Spark 提供了查询结构化数据的能力，查询时既可以使用 SQL 也可以使用人们熟知的 DataFrame API（RDD）。Spark SQL 支持多语言编程包括 Java、Scala、Python 及 R，开发人员可以根据自身喜好进行选择。

DataFrame 是 Spark SQL 的核心，它将数据保存为行构成的集合，行对应列有相应的列名。使用 DataFrames 可以非常方便地查询数据、给数据绘图及进行数据过滤。

DataFrames 也可以用于数据的输入与输出，例如利用 Spark SQL 中的 DataFrames，可以轻易地将下列数据格式加载为表并进行相应的查询操作：

1. RDD；
2. JSON；
3. Hive；
4. Parquet；
5. MySQL；
6. HDFS；

7. S3；

8. JDBC；

9. 其他。

数据一旦被读取，借助于 DataFrames 可以很方便地进行数据过滤、列查询、计数、求平均值及将不同数据源的数据进行整合。

如果你正计划通过读取和写数据来进行分析，Spark SQL 可以轻易地帮你实现并将整个过程自动化。

17.4 实验步骤

登录大数据实验一体机，创建 Spark 集群，并单击搭建 Spark 集群按钮，等待按钮后方的圆点显示为绿色，即搭建完成。平台截图如图 17-1 所示。

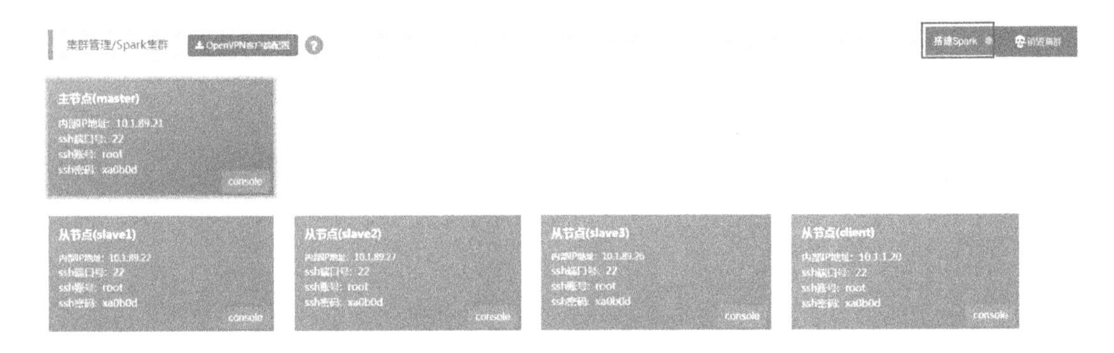

图 17-1 平台截图

在 master 机上建立一个数据文件 weather.dat。

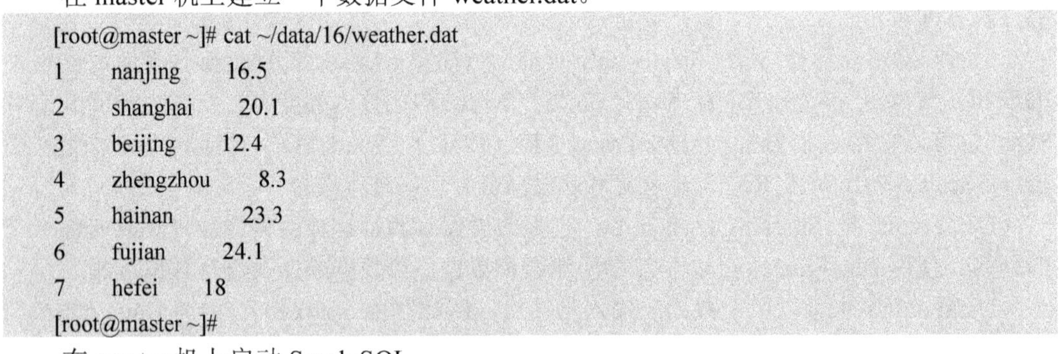

```
[root@master ~]# cat ~/data/16/weather.dat
1	nanjing	16.5
2	shanghai	20.1
3	beijing	12.4
4	zhengzhou	8.3
5	hainan	23.3
6	fujian	24.1
7	hefei	18
[root@master ~]#
```

在 master 机上启动 Spark SQL。

```
[root@master ~]# cd /usr/cstor/spark/
[root@master spark]# bin/spark-sql --master spark://master:7077
```

确认当前 Spark SQL 中是否已经存在我们需要建立的数据库。

```
Spark SQL> show databases;
```

确认在当前的 Spark SQL 中不存在数据库名为 db 的数据库时进行操作。

```
Spark SQL> create database db;
```

切换当前数据库。

```
Spark SQL> use db;
```

建表操作。

```
Spark SQL> create table weather(
id int,
city string,
temperature double
) row format delimited fields terminated by '\t';
```

执行命令检查是否建表成功。

```
Spark SQL> show tables;
```

通过上述命令能在结果中发现 weather 表。

导入数据。

```
Spark SQL> load data local inpath '/root/data/16/weather.dat' overwrite into table weather;
```

执行查询命令。

```
Spark SQL> select * from weather;
Spark SQL> select * from weather where temperature > 10.0;
```

通过查询命令可以正确得到刚才导入的数据就代表导入数据成功。

删除表。

```
[Spark SQL> drop table weather;
```

上述命令可以通过查看数据库中存在的表检查 weather 表是否删除。

17.5　实验结果

建立数据库成功后 show database 结果中能看到以下内容，如图 17-2 所示。

```
16/11/03 15:41:54 INFO scheduler.StatsReportListener: Finished stage: org.apache.spark.scheduler.StageInfo@4d297545
16/11/03 15:41:54 INFO scheduler.DAGScheduler: Job 11 finished: processCmd at CliDriver.java:376, took 0.105746 s
db
default
16/11/03 15:41:54 INFO scheduler.StatsReportListener: task runtime:(count: 1, mean: 84.000000, stdev: 0.000000, max: 84.000000, min:
 84.000000)
16/11/03 15:41:54 INFO scheduler.StatsReportListener:   0%      5%      10%     25%     50%     75%     90%     95%     100%
16/11/03 15:41:54 INFO scheduler.StatsReportListener:   84.0 ms 84.0 ms 84.0 ms 84.0 ms 84.0 ms 84.0 ms 84.0 ms 84.0 ms 84.0 ms
16/11/03 15:41:54 INFO CliDriver: Time taken: 0.165 seconds, Fetched 2 row(s)
spark-sql> 16/11/03 15:41:54 INFO scheduler.StatsReportListener: task result size:(count: 1, mean: 1065.000000, stdev: 0.000000, max
: 1065.000000, min: 1065.000000)
16/11/03 15:41:54 INFO scheduler.StatsReportListener:   0%      5%      10%     25%     50%     75%     90%     95%     100%
16/11/03 15:41:54 INFO scheduler.StatsReportListener:   1065.0 B        1065.0 B        1065.0 B        1065.0 B        1065.0 B   1
065.0 B 1065.0 B        1065.0 B        1065.0 B
16/11/03 15:41:54 INFO scheduler.StatsReportListener: executor (non-fetch) time pct: (count: 1, mean: 0.000000, stdev: 0.000000, max
: 0.000000, min: 0.000000)
16/11/03 15:41:54 INFO scheduler.StatsReportListener:   0%      5%      10%     25%     50%     75%     90%     95%     100%
16/11/03 15:41:54 INFO scheduler.StatsReportListener:    0 %     0 %     0 %     0 %     0 %     0 %     0 %     0 %     0 %
16/11/03 15:41:54 INFO scheduler.StatsReportListener: other time pct: (count: 1, mean: 100.000000, stdev: 0.000000, max: 100.000000,
 min: 100.000000)
16/11/03 15:41:54 INFO scheduler.StatsReportListener:   0%      5%      10%     25%     50%     75%     90%     95%     100%
16/11/03 15:41:54 INFO scheduler.StatsReportListener:   100 %   100 %   100 %   100 %   100 %   100 %   100 %   100 %   100 %
```

图 17-2　执行结果（一）

建表成功之后 show tables 结果中能看到以下内容，如图 17-3 所示。

```
16/11/03 15:42:57 INFO scheduler.TaskSchedulerImpl: Removed TaskSet 12.0, whose tasks have all completed, from pool
16/11/03 15:42:57 INFO scheduler.StatsReportListener: Finished stage: org.apache.spark.scheduler.StageInfo@1555d43c
16/11/03 15:42:57 INFO scheduler.DAGScheduler: Job 12 finished: processCmd at CliDriver.java:376, took 0.119021 s
weather false
16/11/03 15:42:57 INFO CliDriver: Time taken: 0.155 seconds, Fetched 1 row(s)
16/11/03 15:42:57 INFO scheduler.StatsReportListener: task runtime:(count: 1, mean: 98.000000, stdev: 0.000000, max: 98.000000, min:
 98.000000)
spark-sql> 16/11/03 15:42:57 INFO scheduler.StatsReportListener:        0%      5%      10%     25%     50%     75%     90%     95%1
00%
16/11/03 15:42:57 INFO scheduler.StatsReportListener:   98.0 ms 98.0 ms 98.0 ms 98.0 ms 98.0 ms 98.0 ms 98.0 ms 98.0 ms 98.0 ms
16/11/03 15:42:57 INFO scheduler.StatsReportListener: task result size:(count: 1, mean: 1054.000000, stdev: 0.000000, max: 1054.0000
00, min: 1054.000000)
16/11/03 15:42:57 INFO scheduler.StatsReportListener:   0%      5%      10%     25%     50%     75%     90%     95%     100%
16/11/03 15:42:57 INFO scheduler.StatsReportListener:   1054.0 B        1054.0 B        1054.0 B        1054.0 B        1054.0 B   1
054.0 B 1054.0 B        1054.0 B        1054.0 B
16/11/03 15:42:57 INFO scheduler.StatsReportListener: executor (non-fetch) time pct: (count: 1, mean: 0.000000, stdev: 0.000000, max
: 0.000000, min: 0.000000)
16/11/03 15:42:57 INFO scheduler.StatsReportListener:   0%      5%      10%     25%     50%     75%     90%     95%     100%
16/11/03 15:42:57 INFO scheduler.StatsReportListener:    0 %     0 %     0 %     0 %     0 %     0 %     0 %     0 %     0 %
16/11/03 15:42:57 INFO scheduler.StatsReportListener: other time pct: (count: 1, mean: 100.000000, stdev: 0.000000, max: 100.000000,
 min: 100.000000)
16/11/03 15:42:57 INFO scheduler.StatsReportListener:   0%      5%      10%     25%     50%     75%     90%     95%     100%
16/11/03 15:42:57 INFO scheduler.StatsReportListener:   100 %   100 %   100 %   100 %   100 %   100 %   100 %   100 %   100 %
```

图 17-3　执行结果（二）

执行查询命令成功，如图 17-4 所示。

select * from weather;

```
26.0 ms 326.0 ms        326.0 ms        326.0 ms
1       nanjing 16.5
2       shanghai        20.1
3       beijing 12.4
4       zhengzhou       8.3
5       hainan  23.3
6       fujian  24.1
7       hefei   18.0
16/11/03 15:44:28 INFO CliDriver: Time taken: 0.497 seconds, Fetched 7 row(s)
```

图 17-4　执行结果（三）

select * from weather where temperature > 10.0，如图 17-5 所示。

```
16/11/03 15:47:08 INFO scheduler.StatsReportListener:   209.0 ms        209.0 ms        209.0 ms        209.0 ms        225.0 ms        225.0 ms        225.0 ms        225.0 ms        225.0 ms
1       nanjing 16.5
2       shanghai        20.1
3       beijing 12.4
5       hainan  23.3
6       fujian  24.1
7       hefei   18.0
```

图 17-5　执行结果（四）

实验十八　Spark 实验：Spark Streaming

18.1　实验目的

1. 了解 Spark Streaming 版本的 WordCount 和 MapReduce 版本的 WordCount 的区别；

2. 理解 Spark Streaming 的工作流程；

3. 理解 Spark Streaming 的工作原理。

18.2　实验要求

要求实验结束时，每位学生能正确运行本实验中所写的 jar 包程序，能正确地计算出单词数目。

18.3　实验原理

18.3.1　Spark Streaming 架构

计算流程：Spark Streaming 是将流式计算分解成一系列短小的批处理作业。这里的批处理引擎是 Spark，也就是把 Spark Streaming 的输入数据按照 batch size（如 1 秒）分成一段一段的数据（Discretized Stream），每一段数据都转换成 Spark 中的 RDD（Resilient Distributed Dataset），然后将 Spark Streaming 中对 DStream 的 Transformation 操作变为针对 Spark 中对 RDD 的 Transformation 操作，将 RDD 经过操作变成中间结果保存在内存中。整个流式计算根据业务的需求可以对中间的结果进行叠加，或者存储到外部设备，如图 18-1 所示。

容错性：对于流式计算来说，容错性至关重要。首先我们要明确一下 Spark 中 RDD 的容错机制。每一个 RDD 都是一个不可变的分布式可重算的数据集，其记录着确定性的操作继承关系（lineage），所以只要输入数据是可容错的，那么任意一个 RDD 的分区（Partition）出错或不可用，都可以利用原始输入数据通过转换操作而重新算出。

对于 Spark Streaming 来说，其 RDD 的传承关系如图 18-2 所示，图中的每一个椭圆形表示一个 RDD，椭圆形中的每个圆形代表一个 RDD 中的一个 Partition，图中的每一列的多个 RDD 表示一个 DStream（图中有 3 个 DStream），而每一行最后一个 RDD 则表示每一个 Batch Size 所产生的中间结果 RDD。我们可以看到图中的每一个 RDD 都是通过 lineage 相连接的，由于 Spark Streaming 输入数据可以来自于磁盘，例如 HDFS（多份复制）或是来自网络的数据流（Spark Streaming 会将网络输入数据的每一个数据流复制两份到其他的机器）都能保证容错性。所以 RDD 中任意的 Partition 出错，都可以并行地在其他机器上将缺失的 Partition 计算出来。这个容错恢复方式比连续计算模型（如 Storm）的效率更高。

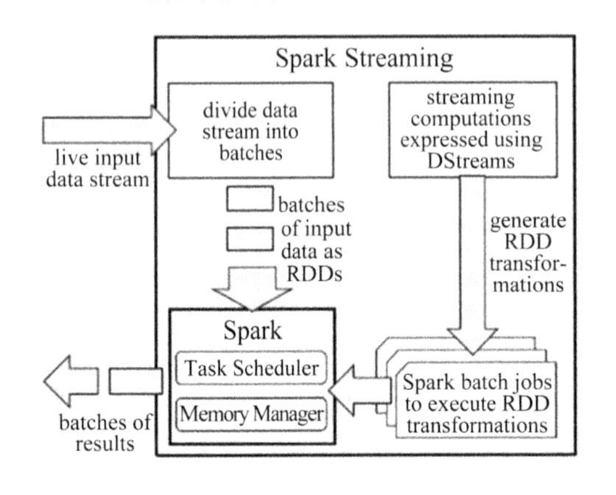

图 18-1　Spark Streaming 计算流程

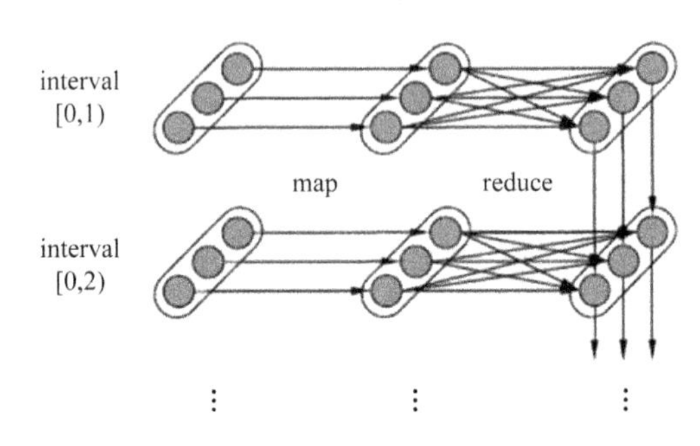

图 18-2　传承 RDD 关系

实时性：对于实时性的讨论，会牵涉流式处理框架的应用场景。Spark Streaming 将流式计算分解成多个 Spark Job，对于每一段数据的处理都会经过 Spark DAG 图分解，以及 Spark 的任务集的调度过程。对于目前版本的 Spark Streaming 而言，其最小的 Batch Size 的选取在 0.5~2 秒（Storm 目前最小的延迟是 100ms 左右），所以 Spark Streaming 能够满足除对实时性要求非常高（如高频实时交易）之外的所有流式准实时计算场景。

扩展性与吞吐量：Spark 目前在 EC2 上已能够线性扩展到 100 个节点（每个节点 4Core），可以以数秒的延迟处理 6GB/s 的数据量（60M records/s），其吞吐量也比流行的 Storm 高 2～5 倍，图 18-3 是 Berkeley 利用 WordCount 和 Grep 两个用例所做的测试，在 Grep 这个测试中，Spark Streaming 中的每个节点的吞吐量是 670k records/s，而 Storm 是 115k records/s。

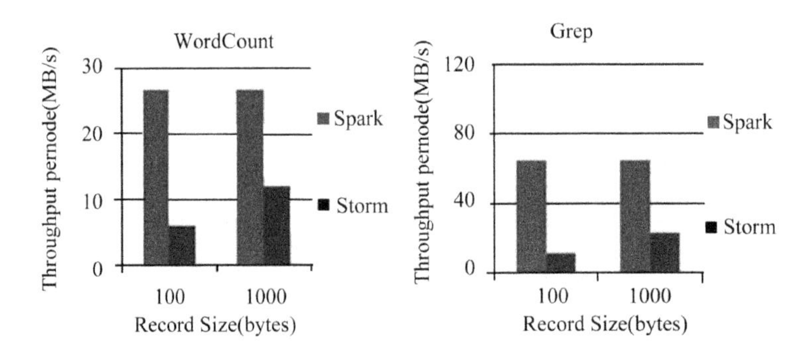

图 18-3　性能比较

18.3.2 Spark Streaming 编程模型

Spark Streaming 的编程和 Spark 的编程如出一辙，对于编程的理解也非常类似。对于 Spark 来说，编程就是对于 RDD 的操作；而对于 Spark Streaming 来说，就是对 DStream 的操作。下面将通过一个大家熟悉的 WordCount 的例子来说明 Spark Streaming 中的输入操作、转换操作和输出操作。

Spark Streaming 初始化：在开始进行 DStream 操作之前，需要对 Spark Streaming 进行初始化生成 StreamingContext。参数中比较重要的是第一个和第三个，第一个参数是指定 Spark Streaming 运行的集群地址，而第三个参数是指定 Spark Streaming 运行时的 batch 窗口大小。在这个例子中就是将 1 秒钟的输入数据进行一次 Spark Job 处理。

```
val ssc = new StreamingContext("Spark://...", "WordCount", Seconds(1), [Homes], [Jars])
```

Spark Streaming 的输入操作：目前 Spark Streaming 已支持了丰富的输入接口，大致分为两类：一类是磁盘输入，如以 batch size 作为时间间隔监控 HDFS 文件系统的某个目录，将目录中内容的变化作为 Spark Streaming 的输入；另一类就是网络流的方式，目前支持 Kafka、Flume、Twitter 和 TCP socket。在 WordCount 例子中，假定通过网络 socket 作为输入流，监听某个特定的端口，最后得出输入 DStream（lines）。

```
val lines = ssc.socketTextStream("localhost",8888)
```

Spark Streaming 的转换操作：与 Spark RDD 的操作极为类似，Spark Streaming 也就是通过转换操作将一个或多个 DStream 转换成新的 DStream。常用的操作包括 map、filter、flatmap 和 join，以及需要进行 shuffle 操作的 groupByKey/reduceByKey 等。在 WordCount 例子中，我们首先需要将 DStream（lines）切分成单词，然后将相同单词的数量进行叠加，最终得到的 WordCounts 就是每一个 batch size 的（单词，数量）中间结果。

```
val words = lines.flatMap(_.split(" "))
val wordCounts = words.map(x => (x, 1)).reduceByKey(_ + _)
```

另外，Spark Streaming 有特定的窗口操作，窗口操作涉及两个参数：一个是滑动窗口的宽度（Window Duration）；另一个是窗口滑动的频率（Slide Duration），这两个参数必须是 batch size 的倍数。例如以过去 5 秒钟为一个输入窗口，每一秒统计一下 WordCount，那么我们会将过去 5 秒钟的每一秒钟的 WordCount 都进行统计，然后进行叠加，得出这个窗口中的单词统计。

```
val wordCounts = words.map(x => (x, 1)).reduceByKeyAndWindow(_ + _, Seconds(5s)，seconds(1))
```

但上面这种方式还不够高效。如果我们以增量的方式来计算就更加高效，例如，计算 t+4 秒这个时刻过去 5 秒窗口的 WordCount，那么我们可以将 t+3 时刻过去 5 秒的统计量加上[t+3，t+4]的统计量，在减去[t−2，t−1]的统计量，这种方法可以复用中间三秒的统计量，提高统计的效率，如图 18-4 所示。

```
val wordCounts = words.map(x => (x, 1)).reduceByKeyAndWindow(_ + _, _ - _, Seconds(5s)，seconds(1))
```

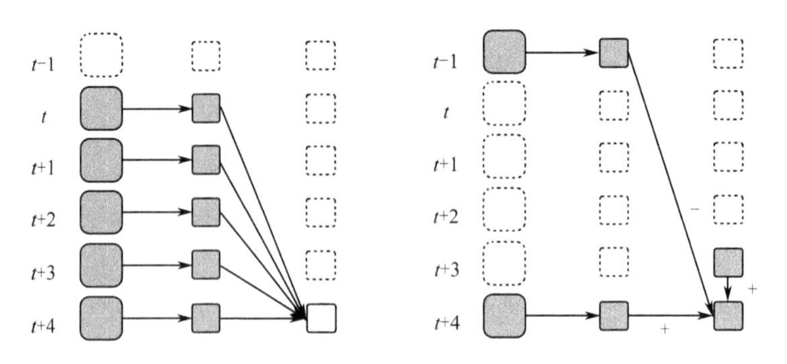

图 18-4　执行过程

Spark Streaming 的输出操作：对于输出操作，Spark 提供了将数据打印到屏幕及输入到文件中。在 WordCount 中我们将 DStream WordCounts 输入到 HDFS 文件中。

```
wordCounts = saveAsHadoopFiles("WordCount")
```

Spark Streaming 的启动：经过上述的操作，Spark Streaming 还没有进行工作，我们还需要调用 Start 操作，Spark Streaming 才开始监听相应的端口，然后收取数据，并进行统计。

```
ssc.start()
```

18.3.3　Spark Streaming 典型案例

在互联网应用中，网站流量统计作为一种常用的应用模式，需要在不同粒度上对不同数据进行统计，既有实时性的需求，又需要涉及聚合、去重、连接等较为复杂的统计需求。传统上，若是使用 Hadoop MapReduce 框架，虽然可以容易地实现较为复杂的统计需求，但实时性却无法得到保证；反之若是采用 Storm 这样的流式框架，实时性虽可以得到保证，但需求的实现复杂度也大大提高了。Spark Streaming 在两者之间找到了一个平衡点，能够以准实时的方式容易地实现较为复杂的统计需求。下面介绍一下使用 Kafka 和 Spark Streaming 搭建实时流量统计框架。

数据暂存：Kafka 作为分布式消息队列，既有非常优秀的吞吐量，又有较高的可靠性和扩展性，在这里采用 Kafka 作为日志传递中间件来接收日志，抓取客户端发送的流量日志，同时接受 Spark Streaming 的请求，将流量日志按序发送给 Spark Streaming 集群。

数据处理：将 Spark Streaming 集群与 Kafka 集群对接，Spark Streaming 从 Kafka 集群中获取流量日志并进行处理。Spark Streaming 会实时地从 Kafka 集群中获取数据并将其存储在内部的可用内存空间中。当每一个 batch 窗口到来时，便对这些数据进行处理。

结果存储：为了便于前端展示和页面请求，处理得到的结果将写入到数据库中。

相比于传统的处理框架，Kafka+Spark Streaming 的架构有以下几个优点。Spark 框架的高效和低延迟保证了 Spark Streaming 操作的准实时性。利用 Spark 框架提供的丰富 API 和高灵活性，可以精简地写出较为复杂的算法。编程模型的高度一致使得上手 Spark Streaming 相当容易，同时也可以保证业务逻辑在实时处理和批处理上的复用。

Spark Streaming 提供了一套高效、可容错的准实时大规模流式处理框架，它能和批处理及即时查询放在同一个软件栈中。如果你学会了 Spark 编程，那么也就学会了 Spark Streaming 编程，如果理解了 Spark 的调度和存储，Spark Streaming 也类似。按照目前的发展趋势，Spark Streaming 一定将会得到更大范围的使用。

18.4　实验步骤

登录大数据实验一体机，创建 Spark 集群，并单击搭建 Spark 集群按钮，等待按钮后方的圆点显示为绿色，即搭建完成。平台截图如图 18-5 所示。

图 18-5　平台截图

使用 jps 检验 Hadoop 集群和 Spark 集群是否成功启动。成功启动 Hadoop 集群和 Spark 集群的情况使用 jps 命令能成功看到以下 Java 进程。

```
[root@master ~]# jps
3711 NameNode
4174 ResourceManager
3957 SecondaryNameNode
4738 Jps
4635 Master
```

打开 IntelliJ IDEA 准备编写 Spark-Steaming 代码。

点击 File→New→Module→Maven→Next→输入 GroupId 和 AriifactId→Next→输入 Module name　新建一个 maven 的 Module。

打开项目录，点击目录下的 pom.xml 文件，在<project>标签中输入 maven 的依赖。然后右键→maven→Reimport 导入 maven 依赖，效果如下。

```
<?xml version="1.0" encoding="UTF-8"?>
<project xmlns="http://maven.apache.org/POM/4.0.0"
          xmlns:xsi="http://www.w3.org/2001/XMLSchema-instance"
          xsi:schemaLocation="http://maven.apache.org/POM/4.0.0
http://maven.apache.org/xsd/maven-4.0.0.xsd">
    <modelVersion>4.0.0</modelVersion>
```

```
    <groupId>com.cstor.sparkstreaming</groupId>
    <artifactId>nice</artifactId>
    <version>1.0-SNAPSHOT</version>
    <build>
        <plugins>
            <plugin>
                <groupId>org.apache.maven.plugins</groupId>
                <artifactId>maven-compiler-plugin</artifactId>
                <configuration>
                    <source>1.6</source>
                    <target>1.6</target>
                </configuration>
            </plugin>
        </plugins>
    </build>


    <!-- https://mvnrepository.com/artifact/org.apache.spark/Spark Streaming_2.10 -->
    <dependencies>
        <dependency>
            <groupId>org.apache.spark</groupId>
            <artifactId>spark-streaming_2.10</artifactId>
            <version>1.6.0</version>
        </dependency>
    </dependencies>

</project>
```

在 src/main/java 的目录下，点击 Java 目录新建一个 package 命名为 spark.streaming.test，然后在包下新建一个 SparkStreaming 的 java class。

在 SparkStreaming 中键入代码。

```
package spark.streaming.test;

import scala.Tuple2;
import com.google.common.collect.Lists;
import org.apache.spark.SparkConf;
import org.apache.spark.api.java.function.FlatMapFunction;
import org.apache.spark.api.java.function.Function2;
import org.apache.spark.api.java.function.PairFunction;
import org.apache.spark.api.java.StorageLevels;
import org.apache.spark.streaming.Durations;
import org.apache.spark.streaming.api.java.JavaDStream;
import org.apache.spark.streaming.api.java.JavaPairDStream;
import org.apache.spark.streaming.api.java.JavaReceiverInputDStream;
```

```
import org.apache.spark.streaming.api.java.JavaStreamingContext;

import java.util.Iterator;
import java.util.regex.Pattern;

public class SparkStreaming {
    private static final Pattern SPACE = Pattern.compile(" ");

    public static void main(String[] args) throws InterruptedException {
        if (args.length < 2) {
            System.err.println("Usage: JavaNetworkWordCount <hostname> <port>");
            System.exit(1);
        }


        SparkConf sparkConf = new SparkConf().setAppName("JavaNetworkWordCount");
        JavaStreamingContext ssc = new JavaStreamingContext(sparkConf, Durations.seconds(1));
        JavaReceiverInputDStream<String> lines = ssc.socketTextStream(
                args[0], Integer.parseInt(args[1]), StorageLevels.MEMORY_AND_DISK_SER);
        JavaDStream<String> words = lines.flatMap(new FlatMapFunction<String, String>() {
            @Override
            public Iterable<String> call(String x){
                return Lists.newArrayList(SPACE.split(x));
            }
        });
        JavaPairDStream<String, Integer> wordCounts = words.mapToPair(
                new PairFunction<String, String, Integer>() {
                    @Override
                    public Tuple2<String, Integer> call(String s) {
                        return new Tuple2<String, Integer>(s, 1);
                    }
                }).reduceByKey(new Function2<Integer, Integer, Integer>() {
            @Override
            public Integer call(Integer i1, Integer i2) {
                return i1 + i2;
            }
        });

        wordCounts.print();
        ssc.start();
        ssc.awaitTermination();
    }
}
```

点击 File→Project Structure→Aritifacts→点击加号→JAR→from modules with dependences→选择刚才新建的 module→选择 Main Class→OK→选择 Output directory→点击 OK。

去掉除 'guava-14.0.1.jar' 和 'guice-3.0.jar'以外所有的 JAR 包，点击 OK。

点击 Build→Build Aritifacts 。选择刚才设置的 jar 包，上传到 master 上去。

新建一个 SSH 连接，登录 master 服务器，使用命令 nc -lk 9999 设置路由器。

```
[root@master ~]# nc -lk 9999
```

注：如果系统只没有 nc 这个命令，可以使用 yum install nc 安装 nc 命令。

进入 Spark 的安装目录，执行下面的命令。

```
[root@master ~]# cd /usr/cstor/spark
[root@master spark]# bin/spark-submit --class spark.streaming.test.SparkStreaming ~/sparkstreaming.jar localhost 9999
```

在网络流中输入单词。按回车结束一次输出。

在命令提交的 xshell 连接中观察程序输出。

18.5 实验结果

在提交任务之后应该能看到以下结果（因屏幕刷新很快，所以只能看到部分结果）。

在 nc -lk 9999 命令下输入，执行结果如图 18-6 所示。

```
[root@master ~]# nc -lk 9999
The weather is nice today The weather is nice today The weather is nice today
```

图 18-6 执行结果（一）

所示结果中应该立刻显示出如图 18-7 所示的内容。

```
-------------------------------------------
Time: 1478259025000 ms
-------------------------------------------
(The,3)
(weather,3)
(is,3)
(nice,3)
(today,3)
```

图 18-7 执行结果（二）

实验十九　Spark 实验：GraphX

19.1　实验目的

1. 了解 Spark 的图计算框架 GraphX 的基本知识；
2. 利用 GraphX 进行建图；
3. 利用 GraphX 进行基本的图操作；
4. 理解 GraphX 图操作的算法。

19.2　实验要求

要求实验结束时，每位学生能正确运行 Spark GraphX 的示例程序，正确上传到集群中运行得到正确的实验结果。并且对实验代码有一定的理解。

19.3　实验原理

Spark GraphX 是一个分布式图处理框架，Spark GraphX 基于 Spark 平台提供对图计算和图挖掘简洁易用又丰富多彩的接口，极大地方便了大家对分布式图处理的需求。

社交网络中人与人之间有很多关系链，例如，Twitter、Facebook、微博、微信，这些都是大数据产生的地方，都需要图计算，现在的图处理基本都是分布式的图处理，而并非单机处理，Spark GraphX 由于底层是基于 Spark 来处理的，所以天然就是一个分布式的图处理系统。

图的分布式或者并行处理其实是把这张图拆分成很多的子图，然后我们分别对这些子图进行计算，计算的时候可以分别迭代进行分阶段的计算，对图进行并行计算。

适用范围：图计算。

19.4　实验步骤

本实验主要可以分为在 IDEA 安装 Scala 的插件、编写 Scala 的程序、打包提交程序和查看程序运行结果等步骤。确认好上述“操作前提”后，读者可按下述步骤执行。

19.4.1　在 Intellij IDEA 中安装 Scala 的插件

注：IntelliJ IDEA 对 Scala 编程比较方便。如不习惯请参考网上 Eclipse 版本。

点击 File→Plugins→输入“Scala”搜索→点击搜索条目为 Scala 的项目→点击右侧绿色按钮 Install→安装完成之后重启 IDEA。

19.4.2 新建 Scala Module

点击当前 IDEA 中的 project 然后右键点击 new→module→选择 Scala→next→输入 module name “sparkgraphx”。

点击新建的 sparkgraphx 的 module 然后右键点击“Add Framework Support”选择 maven 支持。

19.4.3 添加 maven 依赖

在 pom.xml 文件中添加如下依赖。

```
<dependencies>
        <dependency>
            <groupId>org.apache.spark</groupId>
            <artifactId>spark-graphx_2.10</artifactId>
            <version>1.5.1</version>
        </dependency>
</dependencies>
```

右键选项点击 Reimport 导入依赖（需要稍等一段时间导入 maven 依赖）。

19.4.4 新建 Scala 程序

打开项目目录后右键新建 package 命名为 test，然后新建 Scala class 将 Kindt 由 class 改为 object，输入名字为 GraphXExample。

输入代码：

```
package test

import org.apache.log4j.{Level, Logger}
import org.apache.spark.graphx._
import org.apache.spark.rdd.RDD
import org.apache.spark.{SparkConf, SparkContext}

object GraphXExample {
  def main(args: Array[String]) {
    Logger.getLogger("org.apache.spark").setLevel(Level.ERROR)
    Logger.getLogger("org.eclipse.jetty.server").setLevel(Level.OFF)

    val conf = new SparkConf().setAppName("SimpleGraphX").setMaster("local")
    val sc = new SparkContext(conf)

    val vertexArray = Array(
      (1L, ("Alice", 28)),
```

```
      (2L, ("Bob", 27)),
      (3L, ("Charlie", 65)),
      (4L, ("David", 42)),
      (5L, ("Ed", 55)),
      (6L, ("Fran", 50))
    )

    val edgeArray = Array(
      Edge(2L, 1L, 7),
      Edge(2L, 4L, 2),
      Edge(3L, 2L, 4),
      Edge(3L, 6L, 3),
      Edge(4L, 1L, 1),
      Edge(5L, 2L, 2),
      Edge(5L, 3L, 8),
      Edge(5L, 6L, 3)
    )

    val vertexRDD: RDD[(Long, (String, Int))] = sc.parallelize(vertexArray)
    val edgeRDD: RDD[Edge[Int]] = sc.parallelize(edgeArray)

    val graph: Graph[(String, Int), Int] = Graph(vertexRDD, edgeRDD)

    println("属性演示")
    println("**********************************************************")
    println("找出图中年龄大于 30 的顶点：")
    graph.vertices.filter { case (id, (name, age)) => age > 30}.collect.foreach {
      case (id, (name, age)) => println(s"$name is $age")
    }

    println("找出图中属性大于 5 的边：")
    graph.edges.filter(e => e.attr > 5).collect.foreach(e => println(s"${e.srcId} to ${e.dstId} att
${e.attr}"))
    println

    println("列出边属性>5 的tripltes：")
    for (triplet <- graph.triplets.filter(t => t.attr > 5).collect) {
      println(s"${triplet.srcAttr._1} likes ${triplet.dstAttr._1}")
    }
    println

    println("找出图中最大的出度、入度、度数：")
    def max(a: (VertexId, Int), b: (VertexId, Int)): (VertexId, Int) = {
      if (a._2 > b._2) a else b
```

```
        }
        println("max of outDegrees:" + graph.outDegrees.reduce(max) + " max of inDegrees:" + graph.
inDegrees.reduce(max) + " max of Degrees:" + graph.degrees.reduce(max))
        println

        println("**********************************************************")
        println("转换操作")
        println("**********************************************************")
        println("顶点的转换操作，顶点 age + 10：")
        graph.mapVertices{ case (id, (name, age)) => (id, (name, age+10))}.vertices.collect.foreach(v =>
println(s"${v._2._1} is ${v._2._2}"))
        println
        println("边的转换操作，边的属性*2：")
        graph.mapEdges(e=>e.attr*2).edges.collect.foreach(e => println(s"${e.srcId} to ${e.dstId} att
${e.attr}"))
        println

        println("**********************************************************")
        println("结构操作")
        println("**********************************************************")
        println("顶点年纪>30 的子图：")
        val subGraph = graph.subgraph(vpred = (id, vd) => vd._2 >= 30)
        println("子图所有顶点：")
        subGraph.vertices.collect.foreach(v => println(s"${v._2._1} is ${v._2._2}"))
        println
        println("子图所有边：")
        subGraph.edges.collect.foreach(e => println(s"${e.srcId} to ${e.dstId} att ${e.attr}"))
        println


        println("**********************************************************")
        println("连接操作")
        println("**********************************************************")
        val inDegrees: VertexRDD[Int] = graph.inDegrees
        case class User(name: String, age: Int, inDeg: Int, outDeg: Int)

        val initialUserGraph: Graph[User, Int] = graph.mapVertices { case (id, (name, age)) => User(name,
age, 0, 0)}

        val userGraph = initialUserGraph.outerJoinVertices(initialUserGraph.inDegrees) {
          case (id, u, inDegOpt) => User(u.name, u.age, inDegOpt.getOrElse(0), u.outDeg)
        }.outerJoinVertices(initialUserGraph.outDegrees) {
          case (id, u, outDegOpt) => User(u.name, u.age, u.inDeg,outDegOpt.getOrElse(0))
        }
```

```
        println("连接图的属性：")
        userGraph.vertices.collect.foreach(v  =>  println(s"${v._2.name}  inDeg:  ${v._2.inDeg}    outDeg:
${v._2.outDeg}"))
        println

        println("出度和入度相同的人员：")
        userGraph.vertices.filter {
          case (id, u) => u.inDeg == u.outDeg
        }.collect.foreach {
          case (id, property) => println(property.name)
        }
        println


        println("**********************************************************")
        println("聚合操作")
        println("**********************************************************")
        println("找出 5 到各顶点的最短：")
        val sourceId: VertexId = 5L
        val initialGraph = graph.mapVertices((id, _) => if (id == sourceId) 0.0 else Double.PositiveInfinity)
        val sssp = initialGraph.pregel(Double.PositiveInfinity)(
          (id, dist, newDist) => math.min(dist, newDist),
          triplet => {
            if (triplet.srcAttr + triplet.attr < triplet.dstAttr) {
              Iterator((triplet.dstId, triplet.srcAttr + triplet.attr))
            } else {
              Iterator.empty
            }
          },
          (a,b) => math.min(a,b)
        )
        println(sssp.vertices.collect.mkString("\n"))

        sc.stop()
      }
    }
```

19.4.5　程序运行

输入命令运行：

```
[root@client ~]# cd /usr/cstor /spark/
[root@client spark-1.6.0/]# bin/spark-submit --class test.GraphXExample ~/ sparkgraphx.jar
```

19.5 实验结果

程序正确运行结果如图 19-1 所示。

```
16/11/04 16:33:45 INFO Remoting: Starting remoting
16/11/04 16:33:45 INFO Remoting: Remoting started; listening on addresses :[akka.tcp://sparkDriverActorSystem@172.17.0.4:46088]
属性演示
**********************************************************
找出图中年龄大于30的顶点:
David is 42
Fran is 50
Charlie is 65
Ed is 55
找出图中属性大于5的边:
2 to 1 att 7
5 to 3 att 8

列出边属性>5的tripltes:
Bob likes Alice
Ed likes Charlie

找出图中最大的出度、入度、度数:
max of outDegrees:(5,3) max of inDegrees:(2,2) max of Degrees:(2,4)

**********************************************************
转换操作
**********************************************************
顶点的转换操作, 顶点age + 10:
4 is (David,52)
1 is (Alice,38)
6 is (Fran,60)
3 is (Charlie,75)
5 is (Ed,65)
2 is (Bob,37)

边的转换操作, 边的属性*2:
2 to 1 att 14
2 to 4 att 4
3 to 2 att 8
3 to 6 att 6
4 to 1 att 2
5 to 2 att 4
5 to 3 att 16
5 to 6 att 6

**********************************************************
结构操作
**********************************************************
顶点年纪>30的子图:
子图所有顶点:
David is 42
Fran is 50
Charlie is 65
Ed is 55

子图所有边:
3 to 6 att 3
5 to 3 att 8
5 to 6 att 3

**********************************************************
连接操作
**********************************************************
连接图的属性:
David inDeg: 1  outDeg: 1
Alice inDeg: 2  outDeg: 0
Fran inDeg: 2  outDeg: 0
Charlie inDeg: 1  outDeg: 2
Ed inDeg: 0  outDeg: 3
Bob inDeg: 2  outDeg: 2

出度和入度相同的人员:
```

```
出度和入度相同的人员:
David
Bob

**********************************************************
聚合操作
**********************************************************
找出5到各顶点的最短:
(4,4.0)
(1,5.0)
(6,3.0)
(3,8.0)
(5,0.0)
(2,2.0)
16/11/04 16:33:49 INFO RemoteActorRefProvider$RemotingTerminator: Shutting down remote daemon.
[root@client bin]# \
```

图 19-1 执行结果

实验二十　部署 ZooKeeper

20.1　实验目的

掌握 ZooKeeper 集群安装部署，加深对 ZooKeeper 相关概念的理解，熟练 ZooKeeper 的一些常用 Shell 命令。

20.2　实验要求

部署三个节点的 ZooKeeper 集群，通过 ZooKeeper 客户端连接 ZooKeeper 集群，并用 Shell 命令练习创建目录，查询目录等。

20.3　实验原理

ZooKeeper 分布式服务框架是 Apache Hadoop 的一个子项目，它主要是用来解决分布式应用中经常遇到的一些数据管理问题，例如，统一命名服务、状态同步服务、集群管理、分布式应用配置项的管理等。

ZooKeeper 是以 Fast Paxos 算法为基础的。

ZooKeeper 集群的初始化过程：集群中所有机器以投票的方式（少数服从多数）选取某一台机器作为 leader（领导者），其余机器作为 follower（追随者）。如果集群中只有一台机器，那么就这台机器就是 leader，没有 follower。

ZooKeeper 集群与客户端的交互：客户端可以在任意情况下、ZooKeeper 集群中任意一台机器上进行读操作；但是写操作必须得到 leader 的同意后才能执行。

ZooKeeper 选取 leader 的核心算法思想：如果某服务器获得 $N/2 + 1$ 票，则该服务器成为 leader，其中 N 为集群中机器数量。为了避免出现两台服务器获得相同票数（$N/2$），应该确保 N 为奇数，因此构建 ZooKeeper 集群最少需要 3 台机器。

20.4　实验步骤

本实验主要介绍 ZooKeeper 的部署，ZooKeeper 一般部署奇数个节点，部署方法包主要含安装 JDK、修改配置文件、启动测试三个步骤。

20.4.1　安装 JDK

下载安装 JDK，因为 ZooKeeper 服务器在 JVM 上运行。

20.4.2 修改 ZooKeeper 配置文件

首先配置 slave1，slave2，slave3 之间的免密和各个机器的/etc/hosts 文件。

修改 ZooKeeper 的配置文件，步骤如下。

进入解压目录下，把 conf 目录下的 zoo_sample.cfg 复制成 zoo.cfg 文件。

```
cd /usr/cstor/zookeeper/conf
cp zoo_sample.cfg zoo.cfg
```

打开 zoo.cfg 并修改和添加配置项目，如下：

```
# The number of milliseconds of each tick
tickTime=2000
# The number of ticks that the initial
# synchronization phase can take
initLimit=10
# The number of ticks that can pass between
# sending a request and getting an acknowledgement
syncLimit=5
# the port at which the clients will connect
clientPort=2181
# the directory where the snapshot is stored
dataDir=/usr/cstor/zookeeper/data
dataLogDir=/usr/cstor/zookeeper/log
server.1=slave1:2888:3888
server.2=slave2:2888:3888
server.3=slave3:2888:3888
```

新建两个目录。

```
mkdir /usr/cstor/zookeeper/data
mkdir /usr/cstor/zookeeper/log
```

将/usr/cstor/zookeeper 目录传到另外两台机器上。

```
scp -r /usr/cstor/zookeeper root@slave2:/usr/cstor
scp -r /usr/cstor/zookeeper root@slave3:/usr/cstor
```

分别在三个节点上的/usr/local/zookeeper/data 目录下创建一个文件：myid。

```
vi /usr/cstor/zookeeper/data/myid
```

分别在 myid 上按照配置文件的 server.<id>中 id 的数值，在不同机器上的该文件中填写相应过的值，如下：

slave1 的 myid 内容为 1；

slave2 的 myid 内容为 2；

slave3 的 myid 内容为 3。

20.4.3 启动 ZooKeeper 集群

启动 ZooKeeper 集群，进入客户端验证部署完成。

分别在三个节点进入 bin 目录，启动 ZooKeeper 服务进程：

```
cd /usr/cstor/zookeeper/bin
./zkServer.sh start
```

在各机器上依次执行脚本，查看 ZooKeeper 状态信息，两个节点是 follower 状态，一个节点是 leader 状态：

```
./zkServer.sh status
```

在其中一台机器上执行客户端脚本：

```
./zkCli.sh -server slave1:2181,slave2:2181,slave3:2181
```

在客户端 Shell 下执行创建目录命令：

```
create /testZk ""
```

向/testZk 目录写数据：

```
set /testZk 'aaa'
```

读取/testZk 目录数据：

```
get /testZk
```

删除/testZk 目录：

```
rmr /testZk
```

在客户端 Shell 下用 quit 命令退出客户端：

```
quit
```

20.5 实验结果

各个节点执行 jps 命令查看 Java 进程，有 QuorumPeerMain 进程代表该节点 ZooKeeper 安装成功，如图 20-1 所示。

```
16245 Jps
7590 QuorumPeerMain
```

图 20-1 ZooKeeper 进程

在客户端 Shell 下查看 ZooKeeper 集群目录，输入命令：ls /，查看 ZooKeeper 集群目录列表结果如图 20-2～图 20-4 所示。

```
[zk: slave1:2181,slave2:2181,slave3:2181(CONNECTED) 0] ls /
[zookeeper]
[zk: slave1:2181,slave2:2181,slave3:2181(CONNECTED) 1] create /testZk ""
Created /testZk
[zk: slave1:2181,slave2:2181,slave3:2181(CONNECTED) 2] ls /
[testZk, zookeeper]
[zk: slave1:2181,slave2:2181,slave3:2181(CONNECTED) 3]
```

图 20-2 ZooKeeper 客户端结果

```
[zk: localhost:2181(CONNECTED) 5] set /testZk 'aaa'
cZxid = 0x100000002
ctime = Sat Nov 26 13:24:06 UTC 2016
mZxid = 0x1000007fb
mtime = Tue Nov 29 12:56:39 UTC 2016
pZxid = 0x100000002
cversion = 0
dataVersion = 24
aclVersion = 0
ephemeralOwner = 0x0
dataLength = 5
numChildren = 0
```

图 20-3　ZooKeeper 客户端写数据

```
[zk: localhost:2181(CONNECTED) 6] get /testZk
'aaa'
cZxid = 0x100000002
ctime = Sat Nov 26 13:24:06 UTC 2016
mZxid = 0x1000007fb
mtime = Tue Nov 29 12:56:39 UTC 2016
pZxid = 0x100000002
cversion = 0
dataVersion = 24
aclVersion = 0
ephemeralOwner = 0x0
dataLength = 5
numChildren = 0
```

图 20-4　ZooKeeper 客户端读数据

实验二十一　ZooKeeper 进程协作

21.1　实验目的

掌握 Java 代码如何连接 ZooKeeper 集群及通过代码读写 ZooKeeper 集群目录下的数据，掌握 ZooKeeper 如何实现多个线程间的协作。

21.2　实验要求

用 Java 代码实现两个线程，一个向 ZooKeeper 中某一目录中写入数据，另一个监听此目录，若目录下数据有更新则将目录中数据读取并显示出来。

21.3　实验原理

通过 ZooKeeper 实现不同物理机器上的进程间通信。

场景使用：客户端 A 需要向客户端 B 发送一条消息 msg1。

实现方法：客户端 A 把 msg1 发送给 ZooKeeper 集群，然后由客户端 B 自行去 ZooKeeper 集群读取 msg1。

21.4　实验步骤

本实验主要完成多线程通过 ZooKeeper 完成彼此间的协作问题，实验过程包含启动集群、编写代码、客户端提交代码三个步骤。

21.4.1　启动 ZooKeeper 集群

启动 ZooKeeper 集群。具体步骤可以参考实验二十。

21.4.2　导入 jar 包

从 ZooKeeper 安装包的 lib 目录下，将如下 jar 包导入到开发工具。

jline-0.9.94.jar

log4j-1.2.16.jar

netty-3.7.0.Final.jar

slf4j-api-1.6.1.jar

slf4j-log4j12-1.6.1.jar

zookeeper-3.4.6.jar

21.4.3 编写 Java 代码

在开发工具编写 Java 代码，完成实验要求的功能，代码如下。

向/testZk 目录写数据线程代码实现：

```
public class WriteMsg extends Thread {
@Override
public void run() {
        try {
                ZooKeeper zk = new ZooKeeper("slave1:2181", 500000, null);
                String content = Long.toString(new Date().getTime());
                // 修改节点/testZk 下的数据，第三个参数为版本，如果是-1，那会无视被
                   修改的数据版本，直接改掉
                zk.setData("/testZk", content.getBytes(),-1);
                // 关闭 session
                zk.close();
         } catch (Exception e) {
         e.printStackTrace();
      }
   }
}
```

监听/testZk 目录，若数据改变则读取数据并显示线程代码实现：

```
public class ReadMsg {
    public static void main(String[] args) throws Exception {
    final ZooKeeper zk = new ZooKeeper("slave1:2181", 500000, null);
    //定义 watch
    Watcher wacher = new Watcher() {
        public void process(WatchedEvent event) {
        //监听到数据变化取出数据
        if(EventType.NodeDataChanged == event.getType()){
            byte[] bb;
            try {
                bb = zk.getData("/testZk", null, null);
                System.out.println("/testZk 的数据: "+new String(bb));
            } catch (Exception e) {
            e.printStackTrace();
            }
          }
       }
    };
    //设置 watch
    zk.exists("/testZk", wacher);
    //更新/testZk 目录信息，触发 wacth
    while(true)
```

```
        {
            Thread.sleep(2000);
            new WriteMsg().start();
            //watch 一次生效就会删除需重新设置
            zk.exists("/testZk", wacher);
        }
    }
}
```

21.4.4　做成 jar 包

将上述代码打成 jar 包，用工具上传到客户端节点，并执行代码：

```
java -jar ZooKeeperTest.jar
```

21.5　实验结果

在客户端提交 jar 包后运行 Java 代码，当 ZooKeeper 接收线程监控到/testZk 目录信息有变化时，读取该目录的内容，并打印日志，结果如图 21-1 所示。

```
[root@client usr]# java -jar ZooKeeperTest.jar
log4j:WARN No appenders could be found for logger (org.apache.zookeeper.ZooKeeper).
log4j:WARN Please initialize the log4j system properly.
log4j:WARN See http://logging.apache.org/log4j/1.2/faq.html#noconfig for more info.
/testZk的数据: 1477940136996
/testZk的数据: 1477940138997
/testZk的数据: 1477940141000
/testZk的数据: 1477940143004
/testZk的数据: 1477940145007
/testZk的数据: 1477940147011
/testZk的数据: 1477940149014
/testZk的数据: 1477940151017
/testZk的数据: 1477940153020
/testZk的数据: 1477940155023
```

图 21-1　Java 代码运行结果

实验二十二　部署 HBase

22.1　实验目的

1. 掌握 HBase 基础简介及体系架构；
2. 掌握 HBase 集群安装部署及 HBase Shell 一些常用命令的使用；
3. 了解 HBase 和 HDFS 及 ZooKeeper 之间的关系。

22.2　实验要求

1. 巩固学习实验一、实验二、实验二十；
2. 部署一个主节点，三个子节点的 HBase 集群，并引用外部 ZooKeeper；
3. 进入 HBase Shell 通过命令练习创建表、插入数据及查询等命令。

22.3　实验原理

简介：HBase 是基于 Hadoop 的开源分布式数据库，它以 Google 的 BigTable 为原型，设计并实现了具有高可靠性、高性能、列存储、可伸缩、实时读写的分布式数据库系统，它是基于列而不是基于行的模式，适合存储非结构化数据。

体系结构：HBase 是一个分布式的数据库，使用 ZooKeeper 管理集群，使用 HDFS 作为底层存储，它由 HMaster 和 HRegionServer 组成，遵从主从服务器架构。HBase 将逻辑上的表划分成多个数据块即 HRegion，存储在 HRegionServer 中。HMaster 负责管理所有的 HRegionServer，它本身并不存储任何数据，而只是存储数据到 HRegionServer 的映射关系（元数据）。HBase 的基本架构如图 22-1 所示。

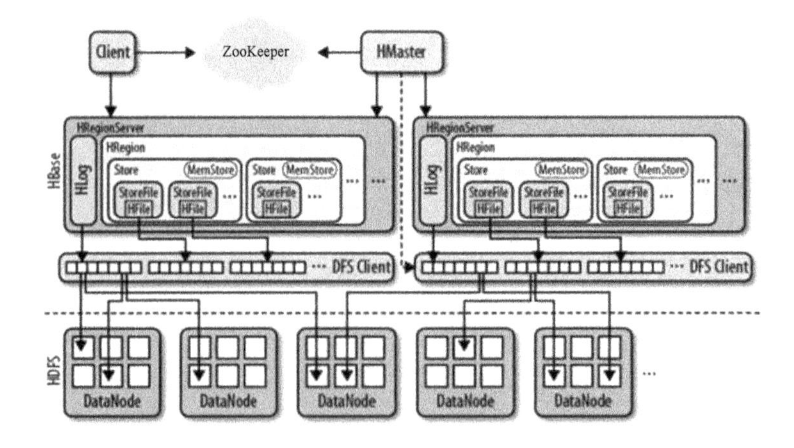

图 22-1　HBase 的基本架构

22.4　实验步骤

本实验主要演示 HBase 的安装部署过程，因 HBase 依赖于 HDFS 和 ZooKeeper，所以该实验需要分为四个步骤。

首先，配置 SSH 无密钥登录（参考实验一）。

其次，安装 Hadoop 集群（参考实验二）。

再次，安装 ZooKeeper 集群（参考实验二十）。

最后，修改 HBase 配置文件，具体内容如下。

将 HBase 安装包 hbase.1.1.2.tar.gz 解压到/usr/cstor 目录，并将 hbase.1.1.2 目录改名为 hbase，且所属用户改成 root:root。

```
[root@master ~]#tar -zxvf hbase.1.1.2.tar.gz -c /usr/cstor/hbase
[root@master ~]#mv /usr/cstor/hbase.1.1.2 /usr/cstor/hbase
[root@master ~]#chown -R root:root /usr/cstor/hbase
```

进入解压目录下，配置 conf 目录下的/usr/cstor/hbase/conf/hbase-env.sh 文件，设置如下：

```
#Java 安装路径
export JAVA_HOME=/usr/local/jdk1.7.0_79 (需根据实际情况指定)
#不使用 HBase 自带的 Zookeeper
export HBASE_MANAGES_ZK=false
```

配置 conf 目录下的 hbase-site.xml 文件，设置如下：

```
<configuration>
    <property>
        <name>hbase.rootdir</name>
        <value>hdfs://master:8020/hbase</value>
    </property>
    <property>
        <name>hbase.cluster.distributed</name>
        <value>true</value>
    </property>
    <property>
        <name>hbase.zookeeper.quorum</name>
        <value>slave1,slave2,slave3</value>
    </property>
    <property>
        <name>hbase.tmp.dir</name>
        <value>/usr/cstor/hbase/data/tmp</value>
    </property>
</configuration>
```

配置 conf 目录下的 regionservers 文件，设置如下：

```
slave1
slave2
slave3
```

配置完成后，将 Hbase 目录传输到集群的其他节点。

```
scp -r /usr/cstor/hbase root@slave1:/usr/cstor
scp -r /usr/cstor/hbase root@slave2:/usr/cstor
scp -r /usr/cstor/hbase root@slave3:/usr/cstor
```

接着，启动 HBase，并简单验证 HBase，如下。

在主节点 master 进入 hbase 解压目录的 bin 目录，启动 HBase 服务进程（已启动 ZooKeeper）：

```
[root@master ~]#cd /usr/cstor/hbase/bin
./start-hbase.sh
```

启动信息如图 22-2 所示。

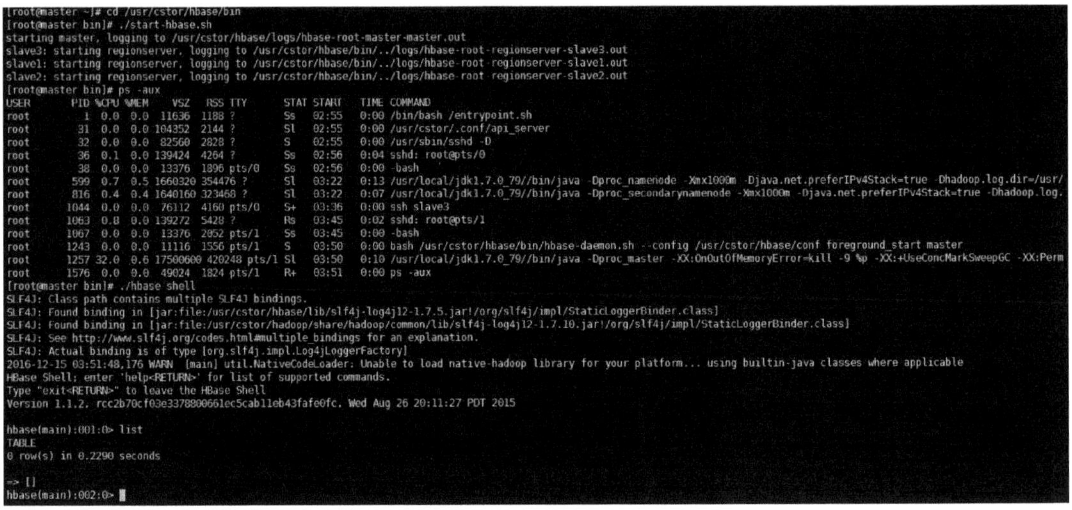

```
[root@master ~]# cd /usr/cstor/hbase/bin
[root@master bin]# ./start-hbase.sh
starting master, logging to /usr/cstor/hbase/logs/hbase-root-master-master.out
slave3: starting regionserver, logging to /usr/cstor/hbase/bin/../logs/hbase-root-regionserver-slave3.out
slave1: starting regionserver, logging to /usr/cstor/hbase/bin/../logs/hbase-root-regionserver-slave1.out
slave2: starting regionserver, logging to /usr/cstor/hbase/bin/../logs/hbase-root-regionserver-slave2.out
[root@master bin]# ps -aux
USER       PID %CPU %MEM    VSZ   RSS TTY      STAT START   TIME COMMAND
root         1  0.0  0.0  11636  1188 ?        Ss   02:55   0:00 /bin/bash /entrypoint.sh
root        31  0.0  0.0 104352  2144 ?        Sl   02:55   0:00 /usr/cstor/.conf/api_server
root        32  0.0  0.0  82560  2828 ?        S    02:55   0:00 /usr/sbin/sshd -D
root        36  0.1  0.0 139424  4264 ?        Ss   02:56   0:04 sshd: root@pts/0
root        38  0.0  0.0  13376  1896 pts/0    Ss   02:56   0:00 -bash
root       599  0.7  0.5 1660320 354476 ?      Sl   03:22   0:13 /usr/local/jdk1.7.0_79//bin/java -Dproc_namenode -Xmx1000m -Djava.net.preferIPv4Stack=true -Dhadoop.log.dir=/usr/
root       816  0.4  0.4 1640160 323468 ?      Sl   03:22   0:07 /usr/local/jdk1.7.0_79//bin/java -Dproc_secondarynamenode -Xmx1000m -Djava.net.preferIPv4Stack=true -Dhadoop.log.
root      1044  0.0  0.0  76112  4160 pts/0    S+   03:36   0:00 ssh slave3
root      1063  0.8  0.0 139272  5428 ?        Rs   03:45   0:02 sshd: root@pts/1
root      1067  0.0  0.0  13376  2052 pts/1    Ss   03:45   0:00 -bash
root      1243  0.0  0.0  11116  1556 pts/1    S    03:50   0:00 bash /usr/cstor/hbase/bin/hbase-daemon.sh --config /usr/cstor/hbase/conf foreground_start master
root      1257 32.0  0.6 17500600 420248 pts/1 Sl   03:50   0:10 /usr/local/jdk1.7.0_79//bin/java -Dproc_master -XX:OnOutOfMemoryError=kill -9 %p -XX:+UseConcMarkSweepGC -XX:Perm
root      1576  0.0  0.0  49024  1824 pts/1    R+   03:51   0:00 ps -aux
[root@master bin]# ./hbase shell
SLF4J: Class path contains multiple SLF4J bindings.
SLF4J: Found binding in [jar:file:/usr/cstor/hbase/lib/slf4j-log4j12-1.7.5.jar!/org/slf4j/impl/StaticLoggerBinder.class]
SLF4J: Found binding in [jar:file:/usr/cstor/hadoop/share/hadoop/common/lib/slf4j-log4j12-1.7.10.jar!/org/slf4j/impl/StaticLoggerBinder.class]
SLF4J: See http://www.slf4j.org/codes.html#multiple_bindings for an explanation.
SLF4J: Actual binding is of type [org.slf4j.impl.Log4jLoggerFactory]
2016-12-15 03:51:48,176 WARN  [main] util.NativeCodeLoader: Unable to load native-hadoop library for your platform... using builtin-java classes where applicable
HBase Shell; enter 'help<RETURN>' for list of supported commands.
Type "exit<RETURN>" to leave the HBase Shell
Version 1.1.2, rcc2b70cf03e3378800661ec5cab11eb43fafe0fc, Wed Aug 26 20:11:27 PDT 2015

hbase(main):001:0> list
TABLE
0 row(s) in 0.2290 seconds

=> []
hbase(main):002:0>
```

图 22-2　启动信息

通过以下命令进入 HBase Shell 界面：

```
./hbase shell
```

在 Shell 里创建表：

```
create 'testhbase' , 'f1'
```

查询所有表名：

```
list
```

查看表结构信息：

```
describe 'testhbase'
```

在 Shell 里插入数据：

```
put 'testhbase', '001', 'f1:name', 'aaa'
```

在 Shell 里查询：

```
scan 'testhbase'
```

删除表，先 disable 再 drop：

```
disable 'testhbase'
drop 'testhbase'
```

22.5 实验结果

HBase 启动成功后，进入 Shell 界面，用 Shell 命令简单操作 HBase 数据库验证 HBase 成功安装，验证结果如图 22-3 所示。

```
hbase(main):006:0> create 'testhbase' , 'f1'
0 row(s) in 1.3050 seconds

=> Hbase::Table - testhbase
hbase(main):007:0> list
TABLE
mytable
testhbase
2 row(s) in 0.0160 seconds

=> ["mytable", "testhbase"]
hbase(main):008:0> describe 'testhbase'
Table testhbase is ENABLED
testhbase
COLUMN FAMILIES DESCRIPTION
{NAME => 'f1', DATA_BLOCK_ENCODING => 'NONE', BLOOMFILTER => 'ROW', REPLICATION_SCOPE => '0', VERSIONS => '1', COMPRESSION => 'NONE', MIN_VERSION
S => '0', TTL => 'FOREVER', KEEP_DELETED_CELLS => 'FALSE', BLOCKSIZE => '65536', IN_MEMORY => 'false', BLOCKCACHE => 'true'}
1 row(s) in 0.0330 seconds

hbase(main):009:0> put 'testhbase', '001', 'f1:name', 'aaa'
0 row(s) in 0.1320 seconds

hbase(main):010:0> scan 'testhbase'
ROW                                     COLUMN+CELL
 001                                    column=f1:name, timestamp=1480437520873, value=aaa
1 row(s) in 0.0480 seconds

hbase(main):011:0> disable 'testhbase'
0 row(s) in 2.2650 seconds

hbase(main):012:0> drop 'testhbase'
0 row(s) in 1.2620 seconds
```

图 22-3 HBase Shell 界面图

HBase 安装成功后，可以通过访问 HBase Web 页面（http://master:16010）来查看 HBase 集群的一些基本情况，如图 22-4 所示。

master:16010/master-status

APACHE HBASE   Home   Table Details   Local Logs   Log Level   Debug Dump   Metrics Dump   HBase Configuration

Master master

Region Servers

Base Stats   Memory   Requests   Storefiles   Compactions

ServerName	Start time	Requests Per Second	Num. Regions
slave1,16020,1478180566307	Thu Nov 03 13:42:48 UTC 2016	0	0
slave2,16020,1478180568457	Thu Nov 03 13:42:48 UTC 2016	0	0
slave3,16020,1478180568282	Thu Nov 03 13:42:48 UTC 2016	0	2
Total:3		0	2

Backup Masters

ServerName	Port	Start Time
Total:0		

Tables

User Tables   System Tables   Snapshots

Tasks

Show All Monitored Tasks   Show non-RPC Tasks   Show All RPC Handler Tasks   Show Active RPC Calls   Show Client Operations   View as JSON

图 22-4 HBase 管理界面

实验二十三　新建 HBase 表

23.1　实验目的

1. 掌握 HBase 数据模型（逻辑模型及物理模型）；

2. 掌握如何使用 Java 代码获得 HBase 连接，并熟练 Java 对 HBase 数据库的基本操作，进一步加深对 HBase 表概念的理解。

23.2　实验要求

通过 Java 代码实现与 HBase 数据库连接，然后用 Java API 创建 HBase 表，向创建的表中写数据，最后将表中数据读取出来并展示。

23.3　实验原理

逻辑模型：HBase 以表的形式存储数据，每个表由行和列组成，每个列属于一个特定的列族（Column Family）。表中的行和列确定的存储单元称为一个元素（Cell），每个元素保存了同一份数据的多个版本，由时间戳（Time Stamp）来标识。行键是数据行在表中的唯一标识，并作为检索记录的主键。在 HBase 中访问表中的行只有三种方式：通过单个行键访问、给定行键的范围扫描、全表扫描。行键可以是任意字符串，默认按字段顺序存储。表中的列定义为: <family>:<qualifier>（<列族>:<限定符>），通过列族和限定符两部分可以唯一指定一个数据的存储列。元素由行键、列（<列族>:<限定符>）和时间戳唯一确定，元素中的数据以字节码的形式存储，没有类型之分。

物理模型：HBase 是按照列存储的稀疏行/列矩阵，其物理模型实际上就是把概念模型中的一个行进行分割，并按照列族存储。

23.4　实验步骤

本实验主要演示 HBase Java API 的一些基本操作，包括取得链接，创建表，写数据，查询等步骤，具体内容如下。

首先，启动 HBase 集群（参考实验二十二）。

其次，从 HBase 安装包的 lib 目录导入如下 jar 包到开发工具（jar 包的版本号以实际的安装中的版本号为主）：

commons-codec-1.4.jar

commons-collections-3.2.2.jar

commons-configuration-1.6.jar

commons-lang-2.6.jar
commons-logging-1.2.jar
guava-12.0.1.jar
hadoop-auth-2.7.2.jar
hadoop-common-2.7.2.jar
hadoop-hdfs-2.7.2.jar
hbase-client-1.1.2.jar
hbase-common-1.1.2.jar
hbase-protocol-1.1.2.jar
htrace-core-3.1.0-incubating.jar
httpclient-4.4.jar
httpcore-4.4.jar
libfb303-0.9.2.jar
log4j-1.2.17.jar
metrics-core-2.2.0.jar
netty-all-4.0.23.Final.jar
protobuf-java-2.5.0.jar
slf4j-api-1.7.7.jar
slf4j-log4j12-1.6.4.jar
zookeeper-3.4.6.jar

然后，获得 HBase 连接，代码实现：

```
Configuration configuration = HBaseConfiguration.create();
Connection connection;
configuration.set("hbase.zookeeper.quorum", "slave1:2181,slave2:2181,slave3:2181");
configuration.set("zookeeper.znode.parent", "/hbase");
connection = ConnectionFactory.createConnection(configuration);
```

然后，通过连接实现对 HBase 数据库的一些基本操作，如下。

新建 HBase 表，代码实现：

```
//获得 HBaseAdmin 对象
Admin admin = connection.getAdmin();
//表名称
String tn = "mytable";
TableName tableName = TableName.valueOf(tn);
//表不存在时创建表
if(!admin.tableExists(tableName))
{
    //创建表描述对象
    HTableDescriptor tableDescriptor = new HTableDescriptor(tableName);
    //列簇 1
    HColumnDescriptor columnDescriptor1 = new HColumnDescriptor("c1".getBytes());
```

```
        tableDescriptor.addFamily(columnDescriptor1);
        //列簇 2
        HColumnDescriptor columnDescriptor2 = new HColumnDescriptor("c2".getBytes());
        tableDescriptor.addFamily(columnDescriptor2);
        //用 HBaseAdmin 对象创建表
        admin.createTable(tableDescriptor);
    }
    //关闭 HBaseAdmin 对象
    admin.close();
```

向表 put 数据，代码实现：

```
    //获得 table 接口
    Table table = connection.getTable(TableName.valueOf("mytable"));
    //添加的数据对象集合
    List<Put> putList = new ArrayList<Put>();
    //添加 10 行数据
    for(int i=0; i<10; i++)
    {
        //put 对象(rowkey)
        String rowkey = "mykey" + i;
        Put put = new Put(rowkey.getBytes());
        //列簇 ，列名，值
        put.addColumn("c1".getBytes(), "c1tofamily1".getBytes(), ("aaa"+i).getBytes());
        put.addColumn("c1".getBytes(), "c2tofamily1".getBytes(), ("bbb"+i).getBytes());
        put.addColumn("c2".getBytes(), "c1tofamily2".getBytes(), ("ccc"+i).getBytes());
        putList.add(put);
    }
    table.put(putList);
    table.close();
```

查询数据，代码实现：

```
    //获得 table 接口（这行代码注意取舍，如果查询的代码和插入代码在同一个类中，则可以不要下
面的这行）
    Table table = connection.getTable(TableName.valueOf("mytable"));
    //Scan 对象
    Scan scan = new Scan();
    //限定 rowkey 查询范围
    scan.setStartRow("mykey0".getBytes());
    scan.setStopRow("mykey9".getBytes());
    //只查询 c1：c1tofamily1 列
    scan.addColumn("c1".getBytes(), "c1tofamily1".getBytes());
    //过滤器集合
    FilterList filterList = new FilterList();
    //查询符合条件 c1：c1tofamily1==aaa7 的记录
    Filter filter1 = new SingleColumnValueFilter("c1".getBytes(), "c1tofamily1".getBytes(),
    CompareFilter.CompareOp.EQUAL, "aaa7".getBytes());
```

```
filterList.addFilter(filter1);
scan.setFilter(filterList);
ResultScanner results = table.getScanner(scan);
for (Result result : results) {
    System.out.println("获得到 rowkey:" + new String(result.getRow()));
    for (Cell cell : result.rawCells()) {
        System.out.println("列簇：" +
        Bytes.toString(cell.getFamilyArray(),cell.getFamilyOffset(),cell.getFamilyLength())
        + "列:" +
  Bytes.toString(cell.getQualifierArray(),cell.getQualifierOffset(),cell.getQualifierLength())
        + "值:" +
        Bytes.toString(cell.getValueArray(), cell.getValueOffset(), cell.getValueLength()));
    }
}
results.close();
table.close();
```

最后，将 Java 代码打成 jar 包，并上传到客户端执行。

```
java -jar HbaseTest.jar
```

附件：完整代码如下。

```
import org.apache.hadoop.conf.Configuration;
import org.apache.hadoop.hbase.*;
import org.apache.hadoop.hbase.client.*;
import org.apache.hadoop.hbase.filter.CompareFilter;
import org.apache.hadoop.hbase.filter.Filter;
import org.apache.hadoop.hbase.filter.FilterList;
import org.apache.hadoop.hbase.filter.SingleColumnValueFilter;
import org.apache.hadoop.hbase.util.Bytes;
import java.io.IOException;
import java.util.ArrayList;
import java.util.List;
public class Main {
public static void main(String[] args) {

        Configuration configuration = HBaseConfiguration.create();
        Connection connection;
        configuration.set("hbase.zookeeper.quorum", "slave1:2181,slave2:2181,slave3:2181");
        configuration.set("zookeeper.znode.parent", "/hbase");
        try {
            connection = ConnectionFactory.createConnection(configuration);
            //获得 HBaseAdmin 对象
            Admin admin = connection.getAdmin();
            //表名称
```

```
        String tn = "mytable";
        TableName tableName = TableName.valueOf(tn);
        //表不存在时创建表
        if (!admin.tableExists(tableName)) {
            //创建表描述对象
            HTableDescriptor tableDescriptor = new HTableDescriptor(tableName);
            //列簇 1
            HColumnDescriptor columnDescriptor1 = new HColumnDescriptor("c1".getBytes());
            tableDescriptor.addFamily(columnDescriptor1);
            //列簇 2
            HColumnDescriptor columnDescriptor2 = new HColumnDescriptor("c2".getBytes());
            tableDescriptor.addFamily(columnDescriptor2);
            //用 HBaseAdmin 对象创建表
            admin.createTable(tableDescriptor);
        }
        //关闭 HBaseAdmin 对象
        admin.close();
        //向表 put 数据，代码实现：
        //获得 table 接口
        Table table = connection.getTable(TableName.valueOf("mytable"));
        //添加的数据对象集合
        List<Put> putList = new ArrayList<Put>();
        //添加 10 行数据
        for (int i = 0; i < 10; i++) {
            //put 对象(rowkey)
            String rowkey = "mykey" + i;
            Put put = new Put(rowkey.getBytes());
            //列簇 ，列名，值
            put.addColumn("c1".getBytes(), "c1tofamily1".getBytes(), ("aaa" + i).getBytes());
            put.addColumn("c1".getBytes(), "c2tofamily1".getBytes(), ("bbb" + i).getBytes());
            put.addColumn("c2".getBytes(), "c1tofamily2".getBytes(), ("ccc" + i).getBytes());
            putList.add(put);
        }
        table.put(putList);
        table.close();
        //查询数据，代码实现：
        //获得 table 接口
//          Table table = connection.getTable(TableName.valueOf("mytable"));
        //Scan 对象
        Scan scan = new Scan();
        //限定 rowkey 查询范围
```

```
                scan.setStartRow("mykey0".getBytes());
                scan.setStopRow("mykey9".getBytes());
                //只查询 c1：c1tofamily1 列
                scan.addColumn("c1".getBytes(), "c1tofamily1".getBytes());
                //过滤器集合
                FilterList filterList = new FilterList();
                //查询符合条件 c1：c1tofamily1==aaa7 的记录
                Filter filter1 = new SingleColumnValueFilter("c1".getBytes(), "c1tofamily1".getBytes(),
CompareFilter.CompareOp.EQUAL, "aaa7".getBytes());
                filterList.addFilter(filter1);
                scan.setFilter(filterList);
                ResultScanner results = table.getScanner(scan);
                for (Result result : results) {
                    System.out.println("获得到 rowkey:" + new String(result.getRow()));
                    for (Cell cell : result.rawCells()) {
                        System.out.println("列簇：" +
                       Bytes.toString(cell.getFamilyArray(), cell.getFamilyOffset(), cell.getFamilyLength())
                            + "列:" +
                      Bytes.toString(cell.getQualifierArray(), cell.getQualifierOffset(), cell.getQualifierLength())
                                + "值:" +
                     Bytes.toString(cell.getValueArray(), cell.getValueOffset(), cell.getValueLength()));
                    }
                }
                results.close();
                table.close();

            } catch (IOException e) {
                e.printStackTrace();
            }
        }
    }
```

23.5　实验结果

表创建完，然后添加数据后，可以通过 shell 查看 mytable 表数据，共插入 10 条数据，数据内容如图 23-1 所示。

客户端提交 jar 包后，运行 Java 代码打印的日志信息如图 23-2 所示（日志中可能会出现乱码，这个没有关系，不影响实验结果的查看）。

```
hbase(main):002:0> scan 'mytable'
ROW                          COLUMN+CELL
 mykey0                      column=c1:c1tofamily1, timestamp=1478000680288, value=aaa0
 mykey0                      column=c1:c2tofamily1, timestamp=1478000680288, value=bbb0
 mykey0                      column=c2:c1tofamily2, timestamp=1478000680288, value=ccc0
 mykey1                      column=c1:c1tofamily1, timestamp=1478000680288, value=aaa1
 mykey1                      column=c1:c2tofamily1, timestamp=1478000680288, value=bbb1
 mykey1                      column=c2:c1tofamily2, timestamp=1478000680288, value=ccc1
 mykey2                      column=c1:c1tofamily1, timestamp=1478000680288, value=aaa2
 mykey2                      column=c1:c2tofamily1, timestamp=1478000680288, value=bbb2
 mykey2                      column=c2:c1tofamily2, timestamp=1478000680288, value=ccc2
 mykey3                      column=c1:c1tofamily1, timestamp=1478000680288, value=aaa3
 mykey3                      column=c1:c2tofamily1, timestamp=1478000680288, value=bbb3
 mykey3                      column=c2:c1tofamily2, timestamp=1478000680288, value=ccc3
 mykey4                      column=c1:c1tofamily1, timestamp=1478000680288, value=aaa4
 mykey4                      column=c1:c2tofamily1, timestamp=1478000680288, value=bbb4
 mykey4                      column=c2:c1tofamily2, timestamp=1478000680288, value=ccc4
 mykey5                      column=c1:c1tofamily1, timestamp=1478000680288, value=aaa5
 mykey5                      column=c1:c2tofamily1, timestamp=1478000680288, value=bbb5
 mykey5                      column=c2:c1tofamily2, timestamp=1478000680288, value=ccc5
 mykey6                      column=c1:c1tofamily1, timestamp=1478000680288, value=aaa6
 mykey6                      column=c1:c2tofamily1, timestamp=1478000680288, value=bbb6
 mykey6                      column=c2:c1tofamily2, timestamp=1478000680288, value=ccc6
 mykey7                      column=c1:c1tofamily1, timestamp=1478000680288, value=aaa7
 mykey7                      column=c1:c2tofamily1, timestamp=1478000680288, value=bbb7
 mykey7                      column=c2:c1tofamily2, timestamp=1478000680288, value=ccc7
 mykey8                      column=c1:c1tofamily1, timestamp=1478000680288, value=aaa8
 mykey8                      column=c1:c2tofamily1, timestamp=1478000680288, value=bbb8
 mykey8                      column=c2:c1tofamily2, timestamp=1478000680288, value=ccc8
 mykey9                      column=c1:c1tofamily1, timestamp=1478000680288, value=aaa9
 mykey9                      column=c1:c2tofamily1, timestamp=1478000680288, value=bbb9
 mykey9                      column=c2:c1tofamily2, timestamp=1478000680288, value=ccc9
10 row(s) in 0.3130 seconds
```

图 23-1　HBase Shell 界面中显示的数据内容

```
[root@client usr]# java -jar HbaseTest.jar
log4j:WARN No appenders could be found for logger (org.apache.hadoop.util.Shell).
log4j:WARN Please initialize the log4j system properly.
log4j:WARN See http://logging.apache.org/log4j/1.2/faq.html#noconfig for more info.
mytable表创建成功!!!
成功插入10条数据!!!
获得到rowkey:mykey7
列簇: c1列:c1tofamily1值:aaa7
```

图 23-2　运行 Java 代码打印的日志信息

实验二十四　部署 Storm

24.1　实验目的

1. 掌握 Storm 基础简介及体系架构；
2. 掌握 Storm 集群安装部署；
3. 掌握 Storm 和 ZooKeeper 之间的关系，并加深对 Storm 架构和原理的理解。

24.2　实验要求

1. 巩固学习实验一、实验二十；
2. 部署四个节点的 Storm 集群，以 master 节点作为主节点，其他三个 slave 节点作为从节点，并修改 Storm Web 的端口为 8081，并引用外部 ZooKeeper。

24.3　实验原理

Storm 简介：Storm 是一个分布式的、高容错的基于数据流的实时处理系统，可以简单、可靠地处理大量的数据流。Storm 支持水平扩展，具有高容错性，保证每个消息都会得到处理，而且处理速度很快（在一个小集群中，每个节点每秒可以处理数以百万计的消息），它有以下特点：编程模型简单、可扩展、高可靠性、高容错性、支持多种编程语言、支持本地模式、高效。Storm 有很多使用场景：如实时分析，在线机器学习，持续计算，分布式 RPC，ETL 等。

体系架构：Storm 共有两层体系结构，第一层采用 master/slave 架构，第二层为 DAG 流式处理器，第一层资源管理器主要负责管理集群资源、响应和调度用户任务，第二层流式处理器则实际执行用户任务。

集群资源管理层：Storm 的集群资源管理器采用 master/slave 架构，主节点即控制节点（master node）和从节点即工作节点（worker node）。控制节点上面运行一个叫 Nimbus 后台服务程序，它的作用类似 Hadoop 里面的 JobTracker， Nimbus 负责在集群里面分发代码，分配计算任务给机器，并且监控状态。每一个工作节点上面运行一个叫做 Supervisor 的服务程序。Supervisor 会监听分配给它那台机器的工作，根据需要启动/关闭工作进程 Worker。每一个工作进程执行一个 Topology 的一个子集；一个运行的 Topology 由运行在很多机器上的很多工作进程 Worker 组成。（一个 Supervisor 里面有多个 Worker，一个 Worker 是一个 JVM。可以配置 Worker 的数量，对应的是 conf/storm.yaml 中的 supervisor.slot 的数量）架构图如图 24-1 所示。

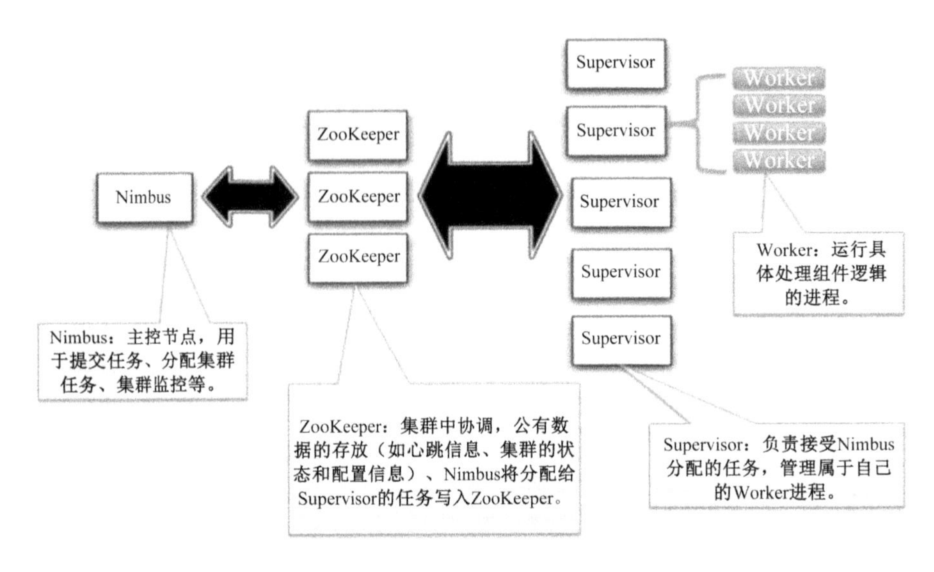

图 24-1 架构图

称集群信息（Nimbus 协议、Supervisor 节点位置）、任务分配信息等关键数据为元数据。Storm 使用 ZooKeeper 集群来共享元数据，这些元数据对 Storm 非常重要，比如 Nimbus 通过这些元数据感知 Supervisor 节点，Supervisor 通过 ZooKeeper 集群感知任务分配情况。Nimbus 和 Supervisor 之间的所有协调工作都是通过 ZooKeeper 集群完成。另外，Nimbus 进程和 Supervisor 进程都是快速失败（fail-fast）和无状态的。所有的状态要么在 ZooKeeper 里面，要么在本地磁盘上。这也就意味着你可以用 kill-9 来杀死 Nimbus 和 Supervisor 进程，然后再重启它们，就好像什么都没有发生过，这个设计使得 Storm 异常地稳定。

数据模型：Storm 实现了一种数据流模型，其中数据持续地流经一个转换实体网络。一个数据流的抽象称为一个流（Stream），这是一个无限的元组序列。元组（Tuple）就像一种使用一些附加的序列化代码来表示标准数据类型（如整数、浮点和字节数组）或用户定义类型的结构。每个流由一个唯一 ID 定义，这个 ID 可用于构建数据源和接收器（Sink）的拓扑结构。流起源于喷嘴（Spout），Spout 将数据从外部来源流入 Storm 拓扑结构中。接收器（或提供转换的实体）称为螺栓（Bolt）。螺栓实现了一个流上的单一转换和一个 Storm 拓扑结构中的所有处理。Bolt 既可实现 MapReduce 之类的传统功能，也可实现更复杂的操作（单步功能），比如过滤、聚合或与数据库等外部实体通信。典型的 Storm 拓扑结构会实现多个转换，因此需要多个具有独立元组流的 Bolt。Bolt 和 Spout 都实现为 Linux 系统中的一个或多个任务。

24.4 实验步骤

本实验主要演示 Storm 集群的安装部署，Storm 依赖于 ZooKeeper，所以该实验大致可分为部署 ZooKeeper、部署 Storm、启动 Storm 集群三个大步骤。

首先，配置 SSH 无密钥登录（参考实验一）。

其次，安装 ZooKeeper 集群（参考实验二十）。

最后，部署 Storm，具体内容如下。

将 Storm 安装包解压到/usr/cstor 目录，并将 Storm 解压目录所属用户改成 root:root。

```
[root@master ~]tar -zxvf apache-storm-0.10.0.tar.gz -c /usr/cstor
[root@master ~]mv /usr/cstor/apache-storm-0.10.0 /usr/cstor/storm
[root@master ~]chown -R root:root /usr/cstor/storm
```

进入解压目录下，把 conf 目录下的 storm.yaml 修改和添加配置项目（每个配置项前面必须留有空格，否则会无法识别），如下。

Storm 集群使用的 ZooKeeper 集群地址，其格式如下：

```
storm.zookeeper.servers:
  - "slave1"
  - "slave2"
  - "slave3"
```

本地存储目录（如果目录不存在记得手动创建: mkdir -p /usr/cstor/storm/workdir)：

```
storm.local.dir: "/usr/cstor/storm/workdir"
```

Storm 集群 Nimbus 机器地址：

```
nimbus.host: "master"
```

对于每个 Supervisor 工作节点，需要配置该工作节点可以运行的 Worker 数量。每个 Worker 占用一个单独的端口用于接收消息，该配置选项即用于定义哪些端口是可被 Worker 使用的。默认情况下，每个节点上可运行 4 个 Workers，分别在 6700、6701、6702 和 6703 端口，如：

```
supervisor.slots.ports:
    - 6700
    - 6701
    - 6702
    - 6703
```

UI 端口（Web 端口，默认 8080）。

```
 ui.port: 8081
```

将/usr/cstor/storm 目录传到另外三个节点上。

```
[root@master ~]scp -r /usr/cstor/storm hadoop@slave1:/usr/cstor
[root@master ~]scp -r /usr/cstor/storm hadoop@slave2:/usr/cstor
[root@master ~]scp -r /usr/cstor/storm hadoop@slave3:/usr/cstor
```

然后，启动 Storm 集群（保证 ZooKeeper 在此之前已启动），启动步骤如下。

主节点（master）启动 Nimbus 服务。

```
cd /usr/cstor/storm/bin
nohup ./storm nimbus >/dev/null 2>&1 &
```

从节点（3 个）启动 Supervisor 服务。

```
cd /usr/cstor/storm/bin
nohup ./storm supervisor >/dev/null 2>&1 &
```

主节点（master）启动 ui 服务。

```
cd /usr/cstor/storm/bin
nohup ./storm ui >/dev/null 2>&1 &
```

24.5 实验结果

在 Storm 集群主节点执行 jps 命令查看 Java 进程，有 nimbus 进程和 core 进程，其中 nimbus 进程为 Storm 主节点进程，core 为 Web 进程，如图 24-2 所示。

在 Storm 集群从节点执行 jps 命令查看 Java 进程，有 supervisor 进程，此进程为 Storm 从节点的进程，如图 24-3 所示。

```
14641 nimbus
15025 core
19859 Jps
```

图 24-2　Storm 主节点进程

```
14721 supervisor
15717 Jps
14409 drpc
15020 logviewer
```

图 24-3　Storm 从节点进程

启动好 Storm 集群后，可以通过流浪器访问 Storm Web 页面（http://master:8081），查看 Storm 集群的一些基本情况，如图 24-4 所示。

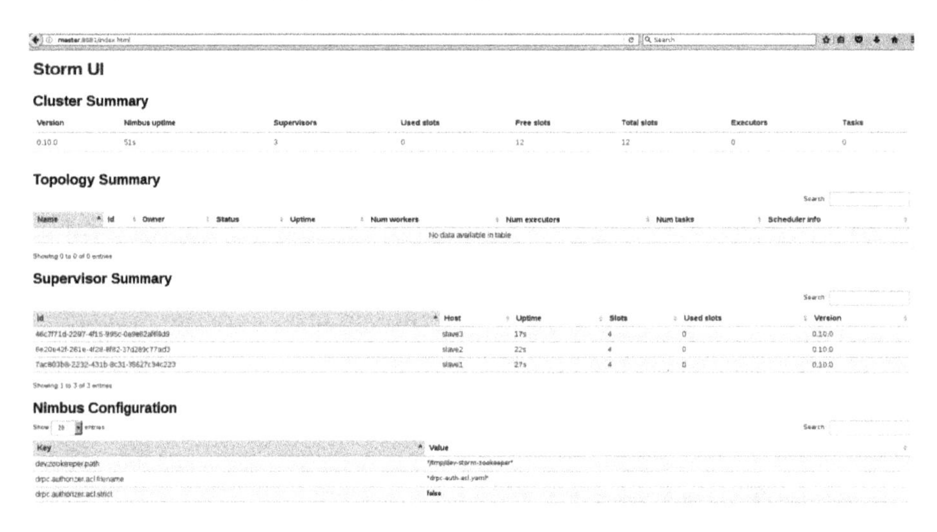

图 24-4　Storm Web 页面

实验二十五　实时 WordCountTopology

25.1　实验目的

掌握用 Java 代码来实现 Storm 任务的拓扑，掌握一个拓扑中 Spout 和 Bolt 的关系及如何组织它们之间的关系，掌握将 Storm 任务提交到集群。

25.2　实验要求

编写一个 Storm 拓扑，一个 Spout 每个一秒钟随机生成一个单词并发射给 Bolt，Bolt 统计接收到的每个单词出现的频率并每隔一秒钟实时打印一次统计结果，最后将任务提交到集群运行，并通过日志查看任务运行结果。

25.3　实验原理

Storm 集群和 Hadoop 集群表面上看很类似。但是 Hadoop 上运行的是 MapReduce jobs，而在 Storm 上运行的是拓扑（Topology），这两者之间是非常不一样的。一个关键的区别是：一个 MapReduce job 最终会结束，而一个 Topology 永远会运行（除非你手动 kill 掉）。

25.3.1　Topologies

一个 Topology 是 Spouts 和 Bolts 组成的图，通过 Stream Groupings 将图中的 Spouts 和 Bolts 连接起来，如图 25-1 所示。

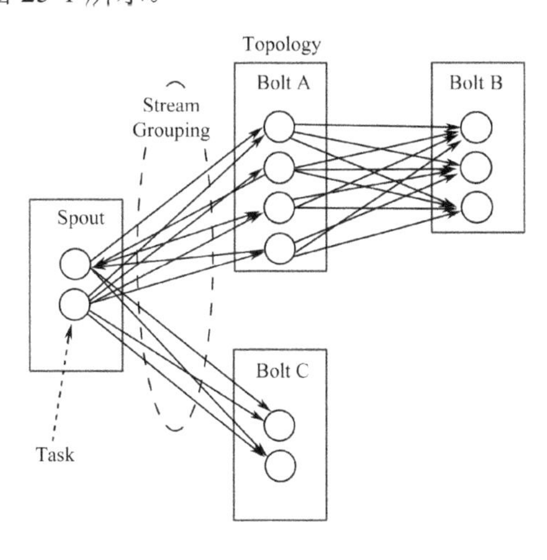

图 25-1　Topology 运行图

一个 Topology 会一直运行直到你手动 kill 掉，Storm 自动重新分配执行失败的任务，并且 Storm 可以保证你不会有数据丢失（如果开启了高可靠性的话）。如果一些机器意外停机，它上面的所有任务会被转移到其他机器上。

运行一个 Topology 很简单。首先，把你所有的代码以及所依赖的 jar 打进一个 jar 包。然后运行类似下面的这个命令：

```
storm jar all-my-code.jar backtype.storm.MyTopology arg1 arg2
```

这个命令会运行主类：backtype.strom.MyTopology，参数是 arg1，arg2。这个类的 main 函数定义这个 Topology 并且把它提交给 Nimbus，Storm jar 负责连接到 Nimbus 并且上传 jar 包。

Topology 的定义是一个 Thrift 结构，并且 Nimbus 就是一个 Thrift 服务， 你可以提交由任何语言创建的 Topology。上面的方面是用 JVM-based 语言提交的最简单的方法。

25.3.2 Spouts

消息源 Spout 是 Storm 里面一个 Topology 里面的消息生产者。一般来说，消息源会从一个外部源读取数据并且向 Topology 里面发出消息：Tuple。Spout 可以是可靠的，也可以是不可靠的。如果这个 Tuple 没有被 Storm 成功处理，可靠的消息源 Spouts 可以重新发射一个 tuple， 但是不可靠的消息源 Spouts 一旦发出一个 Tuple 就不能重发了。

消息源可以发射多条消息流 Stream。使用 OutputFieldsDeclarer.declareStream 来定义多个 Stream，然后使用 SpoutOutputCollector 来发射指定的 stream。

Spout 类里面最重要的方法是 nextTuple。要么发射一个新的 Tuple 到 Topology 里面或者简单的返回如果已经没有新的 Tuple。要注意的是 nextTuple 方法不能阻塞，因为 Storm 在同一个线程上面调用所有消息源 Spout 的方法。

另外两个比较重要的 Spout 方法是 ack 和 fail。Storm 在检测到一个 Tuple 被整个 Topology 成功处理的时候调用 ack，否则调用 fail。Storm 只对可靠的 Spout 调用 ack 和 fail。

25.3.3 Bolts

所有的消息处理逻辑被封装在 Bolts 里面。Bolts 可以做很多事情：过滤，聚合，查询数据库等。

Bolts 可以简单的做消息流的传递。复杂的消息流处理往往需要很多步骤，从而也就需要经过很多 Bolts。例如，算出一堆图片里面被转发最多的图片就至少需要两步：第一步算出每个图片的转发数量，第二步找出转发最多的前 10 个图片。（如果要把这个过程做得更具有扩展性那么可能需要更多的步骤。）

Bolts 可以发射多条消息流，使用 OutputFieldsDeclarer.declareStream 定义 Stream，使用 OutputCollector.emit 来选择要发射的 Stream。

Bolts 的主要方法是 Execute，它以一个 Tuple 作为输入，Bolts 使用 OutputCollector 来发射 Tuple，Bolts 必须要为它处理的每一个 Tuple 调用 OutputCollector 的 ack 方法，以通知 Storm 这个 Tuple 被处理完成了，从而通知这个 Tuple 的发射者 Spouts。一般的

流程是：Bolts 处理一个输入 Tuple，发射 0 个或者多个 Tuple，然后调用 ack 通知 Storm 自己已经处理过这个 Tuple 了。Storm 提供了一个 IBasicBolt 会自动调用 ack。

25.4　实验步骤

本实验主要演示一个完整的 Storm 拓扑编码过程，主要包含 Spout、Bolt 和构建 Topology 几个步骤。

首先，启动 Storm 集群。

其次，将 Storm 安装包的 lib 目录内如下 jar 包导入到开发工具（见图 25-2）。

asm-4.0.jar	2017/2/7 14:53	Executable Jar File	45 KB
clojure-1.6.0.jar	2017/2/7 14:53	Executable Jar File	3,579 KB
disruptor-2.10.4.jar	2017/2/7 14:53	Executable Jar File	52 KB
hadoop-auth-2.4.0.jar	2017/2/7 14:53	Executable Jar File	50 KB
kryo-2.21.jar	2017/2/7 14:53	Executable Jar File	355 KB
log4j-api-2.1.jar	2017/2/7 14:53	Executable Jar File	131 KB
log4j-core-2.1.jar	2017/2/7 14:53	Executable Jar File	806 KB
log4j-over-slf4j-1.6.6.jar	2017/2/7 14:53	Executable Jar File	21 KB
log4j-slf4j-impl-2.1.jar	2017/2/7 14:53	Executable Jar File	23 KB
minlog-1.2.jar	2017/2/7 14:53	Executable Jar File	5 KB
reflectasm-1.07-shaded.jar	2017/2/7 14:53	Executable Jar File	65 KB
servlet-api-2.5.jar	2017/2/7 14:53	Executable Jar File	103 KB
slf4j-api-1.7.7.jar	2017/2/7 14:53	Executable Jar File	29 KB
storm-core-0.10.0.jar	2017/2/7 14:53	Executable Jar File	15,963 KB

图 25-2　相关 jar 包列表

再次，编写代码，实现一个完整的 Topology，内容如下。

Spout 随机发送单词，代码实现：

```
package cproc.word;

import java.util.Map;
import java.util.Random;
import backtype.storm.spout.SpoutOutputCollector;
import backtype.storm.task.TopologyContext;
import backtype.storm.topology.OutputFieldsDeclarer;
import backtype.storm.topology.base.BaseRichSpout;
import backtype.storm.tuple.Fields;
import backtype.storm.tuple.Values;
import backtype.storm.utils.Utils;

public class WordReaderSpout extends BaseRichSpout {
    private SpoutOutputCollector collector;
    @Override
```

```
    public void open(Map conf, TopologyContext context, SpoutOutputCollector collector)
    {
        this.collector = collector;
    }
    @Override
    public void nextTuple() {
            //这个方法会不断被调用，为了降低它对 CPU 的消耗，让它 sleep 一下
     Utils.sleep(1000);
     final String[] words = new String[] {"nathan", "mike", "jackson", "golda", "bertels"};
     Random rand = new Random();
     String word = words[rand.nextInt(words.length)];
     collector.emit(new Values(word));
    }
    @Override
    public void declareOutputFields(OutputFieldsDeclarer declarer) {
        declarer.declare(new Fields("word"));
    }
}
```

Bolt 单词计数，并每隔一秒打印一次，代码实现：

```
package cproc.word;

import java.util.HashMap;
import java.util.Map;
import java.util.Map.Entry;
import backtype.storm.task.TopologyContext;
import backtype.storm.topology.BasicOutputCollector;
import backtype.storm.topology.OutputFieldsDeclarer;
import backtype.storm.topology.base.BaseBasicBolt;
import backtype.storm.tuple.Tuple;

public class WordCounterBolt extends BaseBasicBolt {
    private static final long serialVersionUID = 5683648523524179434L;
    private HashMap<String, Integer> counters = new HashMap<String, Integer>();
    private volatile boolean edit = false;
    @Override
    public void prepare(Map stormConf, TopologyContext context) {
        //定义一个线程 1 秒钟打印一次统计的信息
        new Thread(new Runnable() {
          public void run() {
              while (true) {
                if (edit) {
                    for (Entry<String, Integer> entry : counters.entrySet())
                    {
                        System.out.println(entry.getKey() + " : " + entry.getValue());
```

```
                    }
                    edit = false;
                }
                try {
                    Thread.sleep(1000);
                } catch (InterruptedException e) {
                    e.printStackTrace();
                }
            }
        }
    }).start();
    }
    @Override
    public void execute(Tuple input, BasicOutputCollector collector) {
        String str = input.getString(0);
        if (!counters.containsKey(str)) {
            counters.put(str, 1);
        } else {
            Integer c = counters.get(str) + 1;
            counters.put(str, c);
        }
        edit = true;
    }
    @Override
    public void declareOutputFields(OutputFieldsDeclarer declarer) {

    }
}
```

构建 Topology 并提交到集群主函数，代码实现：

```
package cproc.word;

import backtype.storm.Config;
import backtype.storm.StormSubmitter;
import backtype.storm.generated.AlreadyAliveException;
import backtype.storm.generated.AuthorizationException;
import backtype.storm.generated.InvalidTopologyException;
import backtype.storm.topology.TopologyBuilder;

public class WordCountTopo {
    public static void main(String[] args) throws Exception{
      //构建 Topology
      TopologyBuilder builder = new TopologyBuilder();
      builder.setSpout("word-reader", new WordReaderSpout());
      builder.setBolt("word-counter", new WordCounterBolt())
```

```
        .shuffleGrouping("word-reader");
        Config conf = new Config();
        //集群方式提交
        StormSubmitter.submitTopologyWithProgressBar("wordCount", conf,
        builder.createTopology());
    }
}
```

最后，将 Storm 代码打成 wordCount-Storm.jar（打包的时候不要包含 Storm 中的 jar，不然会报错的，将无法运行，即：wordCount-Storm.jar 中只包含上面三个类的代码）上传到主节点的/usr/cstor/storm/bin 目录下，在主节点进入 Storm 安装目录的 bin 下面用以下命令提交任务：

```
cd /usr/cstor/storm/bin
./storm jar wordCount-Storm.jar cproc.word.WordCountTopo wordCount
```

使用以下命令结束 storm 任务：

```
./storm kill wordCount
```

25.5 实验结果

Storm 任务执行时，可以查看 Storm 日志文件，日志里面打印了统计的单词结果，日志内容如图 25-3 所示。

```
2016-10-18 16:33:53.933 STDIO [INFO] jackson : 87
2016-10-18 16:33:53.933 STDIO [INFO] mike : 91
2016-10-18 16:33:53.933 STDIO [INFO] nathan : 81
2016-10-18 16:33:53.933 STDIO [INFO] golda : 99
2016-10-18 16:33:53.933 STDIO [INFO] bertels : 78
2016-10-18 16:33:54.934 STDIO [INFO] jackson : 88
2016-10-18 16:33:54.934 STDIO [INFO] mike : 91
2016-10-18 16:33:54.934 STDIO [INFO] nathan : 81
2016-10-18 16:33:54.934 STDIO [INFO] golda : 99
2016-10-18 16:33:54.934 STDIO [INFO] bertels : 78
2016-10-18 16:33:55.935 STDIO [INFO] jackson : 88
2016-10-18 16:33:55.935 STDIO [INFO] mike : 91
2016-10-18 16:33:55.935 STDIO [INFO] nathan : 81
```

图 25-3　Storm 任务日志

实验二十六　文件数据 Flume 至 HDFS

26.1　实验目的

1. 掌握 Flume 的安装部署；
2. 掌握一个 Agent 中 Source、Sink、Channel 组件之间的关系；
3. 加深对 Flume 结构和概念的理解；
4. 掌握 Flume 的编码方法及启动任务方法。

26.2　实验要求

1. 在一台机器上（本例以 slave1 为例）部署 Flume；

2. 实时收集本地 Hadoop 日志的最新信息然后将收集到日志信息以一分钟一个文件的形式写入 HDFS 目录中。

26.3　实验原理

Flume 是 Cloudera 提供的一个高可用的，高可靠的，分布式的海量日志采集、聚合和传输的系统，Flume 支持在日志系统中定制各类数据发送方，用于收集数据；同时，Flume 提供对数据进行简单处理，并写到各种数据接受方（可定制）的能力。

Flume 提供对数据进行简单处理，并写到各种数据接受方（可定制）的能力，Flume 提供了从 console（控制台）、RPC（Thrift-RPC）、text（文件）、tail（UNIX tail）、syslog（syslog 日志系统，支持 TCP 和 UDP 两种模式），exec（命令执行）等数据源上收集数据的能力。

当前 Flume 有两个版本，Flume 0.9X 版本的统称 Flume-og，Flume1.X 版本的统称 Flume-ng。由于 Flume-ng 经过重大重构，与 Flume-og 有很大不同，使用时请注意区分。

Flume-og 采用了多 Master 的方式。为了保证配置数据的一致性，Flume 引入了 ZooKeeper，用于保存配置数据，ZooKeeper 本身可保证配置数据的一致性和高可用，另外，在配置数据发生变化时，ZooKeeper 可以通知 Flume Master 节点。Flume Master 间使用 gossip 协议同步数据。

Flume-ng 最明显的改动就是取消了集中管理配置的 Master 和 ZooKeeper，变为一个纯粹的传输工具。Flume-ng 另一个主要的不同点是读入数据和写出数据由不同的工作线程处理（称为 Runner）。在 Flume-og 中，读入线程同样做写出工作（除了故障重试）。如果写出慢的话（不是完全失败），它将阻塞 Flume 接收数据的能力。这种异步的设计使读入线程可以顺畅地工作而无须关注下游的任何问题。

Flume 以 Agent 为最小的独立运行单位。一个 Agent 就是一个 JVM。单 Agent 由 Source、Sink 和 Channel 三大组件构成，如图 26-1 所示。

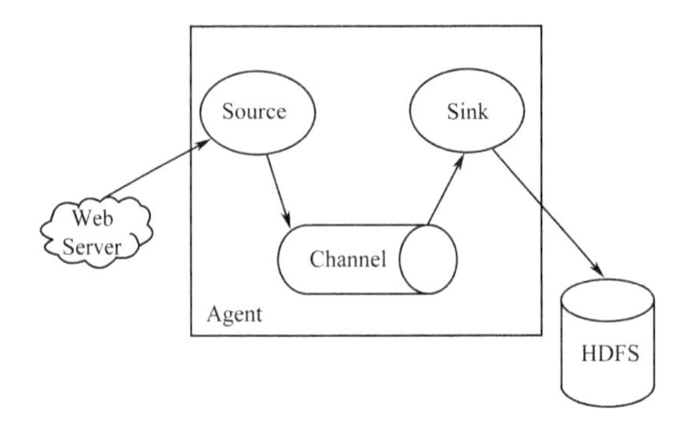

图 26-1 Flume 运行图

值得注意的是，Flume 提供了大量内置的 Source、Channel 和 Sink 类型（见图 26-2）。不同类型的 Source、Channel 和 Sink 可以自由组合。组合方式基于用户设置的配置文件，非常灵活。比如：Channel 可以把事件暂存在内存里，也可以持久化到本地硬盘上。Sink 可以把日志写入 HDFS，HBase，甚至是另外一个 Source，等等。Flume 支持用户建立多级流，也就是说，多个 Agent 可以协同工作，并且支持 Fan-in、Fan-out、Contextual Routing、Backup Routes，这也正是 Flume 厉害之处。

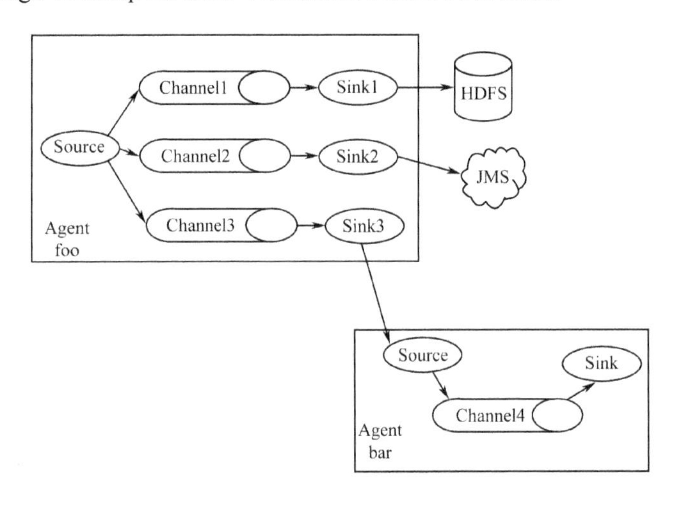

图 26-2 Source、Channel 和 Sink 类型

26.3.1 Flume 的特点

Flume 是一个分布式、可靠和高可用的海量日志采集、聚合和传输的系统。支持在日志系统中定制各类数据发送方，用于收集数据；同时，Flume 提供对数据进行简单处理，并写到各种数据接受方（比如文本、HDFS、Hbase 等）的能力。

Flume 的数据流由事件（Event）贯穿始终。事件是 Flume 的基本数据单位，它携带日志数据（字节数组形式）并且携带有头信息，这些 Event 由 Agent 外部的 Source 生成，当 Source 捕获事件后会进行特定的格式化，然后 Source 会把事件推入（单个或多个）Channel 中。你可以把 Channel 看作是一个缓冲区，它将保存事件直到 Sink 处理完该事件。Sink 负责持久化日志或者把事件推向另一个 Source。

26.3.2　Flume 的可靠性

当节点出现故障时，日志能够被传送到其他节点上而不会丢失。Flume 提供了三种级别的可靠性保障，从强到弱依次分别为：end-to-end（收到数据 agent 首先将 event 写到磁盘上，当数据传送成功后，再删除；如果数据发送失败，可以重新发送），Store on failure（这也是 scribe 采用的策略，当数据接收方 crash 时，将数据写到本地，待恢复后，继续发送），Besteffort（数据发送到接收方后，不会进行确认）。

26.4　实验步骤

本实验主要演示 Flume 安装以及启动一个 Flume 收集日志信息的例子，实验主要包含以下三个大步骤。

首先，启动 Hadoop 集群。

其次，（剩下的所有步骤只需要在 master 上操作就可以了）安装并配置 Flume 任务，内容如下。

将 Flume 安装包解压到/usr/cstor 目录，并将 flume 目录所属用户改成 root:root。

```
tar -zxvf flume-1.5.2.tar.gz -c /usr/cstor
chown -R root:root /usr/cstor/flume
```

进入解压目录下，在 conf 目录下新建 test.conf 文件并添加以下配置内容。

```
#定义 agent 中各组件名称
agent1.sources=source1
agent1.sinks=sink1
agent1.channels=channel1

# source1 组件的配置参数
agent1.sources.source1.type=exec
#此处的文件/home/source.log 需要手动生成，见后续说明
agent1.sources.source1.command=tail -n +0 -F /home/source.log

# channel1 的配置参数
agent1.channels.channel1.type=memory
agent1.channels.channel1.capacity=1000
agent1.channels.channel1.transactionCapactiy=100

# sink1 的配置参数
agent1.sinks.sink1.type=hdfs
agent1.sinks.sink1.hdfs.path=hdfs://master:8020/flume/data
```

```
agent1.sinks.sink1.hdfs.fileType=DataStream
#时间类型
agent1.sinks.sink1.hdfs.useLocalTimeStamp=true
agent1.sinks.sink1.hdfs.writeFormat=TEXT
#文件前缀
agent1.sinks.sink1.hdfs.filePrefix=%Y-%m-%d-%H-%M
#60 秒滚动生成一个文件
agent1.sinks.sink1.hdfs.rollInterval=60
#HDFS 块副本数
agent1.sinks.sink1.hdfs.minBlockReplicas=1
#不根据文件大小滚动文件
agent1.sinks.sink1.hdfs.rollSize=0
#不根据消息条数滚动文件
agent1.sinks.sink1.hdfs.rollCount=0
#不根据多长时间未收到消息滚动文件
agent1.sinks.sink1.hdfs.idleTimeout=0

# 将 source 和 sink 绑定到 channel
agent1.sources.source1.channels=channel1
agent1.sinks.sink1.channel=channel1
```

再次，在 HDFS 上创建/flume/data 目录：

```
cd /usr/cstor/hadoop/bin
./hdfs dfs -mkdir /flume
./hdfs dfs -mkdir /flume/data
```

最后，进入 Flume 安装的 bin 目录下，

```
cd /usr/cstor/flume/bin
```

启动 Flume，开始收集日志信息。

```
[root@master bin]# ./flume-ng agent --conf conf --conf-file /usr/cstor/flume/conf/test.conf --name agent1 -Dflume.root.logger=DEBUG,console
```

看到如下结果就表示启动成功：

```
17/02/08 10:31:40 INFO node.PollingPropertiesFileConfigurationProvider: Configuration provider starting
17/02/08 10:31:40 INFO node.PollingPropertiesFileConfigurationProvider: Reloading configuration file:/usr/cstor/flume/conf/test.conf
17/02/08 10:31:40 INFO conf.FlumeConfiguration: Processing:sink1
17/02/08 10:31:40 INFO conf.FlumeConfiguration: Processing:sink1
17/02/08 10:31:40 INFO conf.FlumeConfiguration: Processing:sink1
17/02/08 10:31:40 INFO conf.FlumeConfiguration: Processing:sink1
17/02/08 10:31:40 INFO conf.FlumeConfiguration: Processing:sink1
17/02/08 10:31:40 INFO conf.FlumeConfiguration: Processing:sink1
17/02/08 10:31:40 INFO conf.FlumeConfiguration: Processing:sink1
17/02/08 10:31:40 INFO conf.FlumeConfiguration: Added sinks: sink1 Agent: agent1
17/02/08 10:31:40 INFO conf.FlumeConfiguration: Processing:sink1
17/02/08 10:31:40 INFO conf.FlumeConfiguration: Processing:sink1
17/02/08 10:31:40 INFO conf.FlumeConfiguration: Processing:sink1
17/02/08 10:31:40 INFO conf.FlumeConfiguration: Processing:sink1
17/02/08 10:31:40 INFO conf.FlumeConfiguration: Processing:sink1
17/02/08 10:31:40 INFO conf.FlumeConfiguration: Post-validation flume configuration contains configuration for agents: [agent1]
17/02/08 10:31:40 INFO node.AbstractConfigurationProvider: Creating channels
17/02/08 10:31:40 INFO channel.DefaultChannelFactory: Creating instance of channel channel1 type memory
17/02/08 10:31:40 INFO node.AbstractConfigurationProvider: Created channel channel1
17/02/08 10:31:40 INFO source.DefaultSourceFactory: Creating instance of source source1, type exec
17/02/08 10:31:40 INFO sink.DefaultSinkFactory: Creating instance of sink: sink1, type: hdfs
17/02/08 10:31:40 WARN util.NativeCodeLoader: Unable to load native-hadoop library for your platform... using builtin-java classes where applicable
17/02/08 10:31:41 INFO hdfs.HDFSEventSink: Hadoop Security enabled: false
17/02/08 10:31:41 INFO node.AbstractConfigurationProvider: Channel channel1 connected to [source1, sink1]
17/02/08 10:31:41 INFO node.Application: Starting new configuration:{ sourceRunners:{source1=EventDrivenSourceRunner: { source:org.apache.flume.source.ExecSource{name:source1,state:IDLE} }} sinkRunners:{sink1
or@609acf2c counterGroup:{ name:null counters:{} } }} channels:{channel1=org.apache.flume.channel.MemoryChannel{name: channel1}} }
17/02/08 10:31:41 INFO node.Application: Starting Channel channel1
17/02/08 10:31:41 INFO instrumentation.MonitoredCounterGroup: Monitored counter group for type: CHANNEL, name: channel1: Successfully registered new MBean.
17/02/08 10:31:41 INFO instrumentation.MonitoredCounterGroup: Component type: CHANNEL, name: channel1 started
17/02/08 10:31:41 INFO node.Application: Starting Sink sink1
17/02/08 10:31:41 INFO node.Application: Starting Source source1
17/02/08 10:31:41 INFO source.ExecSource: Exec source starting with command:tail -n +0 -F /usr/cstor/hadoop/logs/hadoop-hadoop-datanode-slave1.log
17/02/08 10:31:41 INFO instrumentation.MonitoredCounterGroup: Monitored counter group for type: SINK, name: sink1: Successfully registered new MBean.
17/02/08 10:31:41 INFO instrumentation.MonitoredCounterGroup: Component type: SINK, name: sink1 started
17/02/08 10:31:41 INFO instrumentation.MonitoredCounterGroup: Monitored counter group for type: SOURCE, name: source1: Successfully registered new MBean.
17/02/08 10:31:41 INFO instrumentation.MonitoredCounterGroup: Component type: SOURCE, name: source1 started
```

启动成功之后需要手动生成消息源即配置文件中的/home/source.log，使用如下命令去不断的写入文字到/home/source.log 中：

```
[root@master bin]# vi /home/source.log
[root@master bin]# echo aa >> /home/source.log
[root@master bin]# echo aa >> /home/source.log
[root@master bin]# echo aa >> /home/source.log
[root@master bin]# echo aa >> /home/source.log
[root@master bin]# echo aa >> /home/source.log
[root@master bin]# echo aa >> /home/source.log
[root@master bin]# echo aa >> /home/source.log
[root@master bin]# echo aa >> /home/source.log
[root@master bin]# echo aa >> /home/source.log
[root@master bin]# cat /home/source.log
aa
aa
aa
aa
aa
aa
aa
aa
aa
```


到此就可以去查看实验结果了。

26.5　实验结果

实验结果查看的步骤：

```
[root@master bin]# cd /usr/cstor/hadoop/bin
[root@master bin]# hadoop fs -ls /flume/data
17/02/08 10:56:11 WARN util.NativeCodeLoader: Unable to load native-hadoop library for your platform... using builtin-java classes where applicable
Found 2 items
-rw-r--r--   3 root supergroup         15 2017-02-08 10:48 /flume/data/2017-02-08-10-47.1486522051447
-rw-r--r--   3 root supergroup         36 2017-02-08 10:49 /flume/data/2017-02-08-10-48.1486522082469
```

实验二十七 Kafka 订阅推送示例

27.1 实验目的

1. 掌握 Kafka 的安装部署；
2. 掌握 Kafka 的 Topic 创建及如何生成消息和消费消息；
3. 掌握 Kafka 和 ZooKeeper 之间的关系；
4. 了解 Kafka 如何保存数据及加深对 Kafka 相关概念的理解。

27.2 实验要求

在三台机器上（以 slave1、slave2、slave3 为例），分别部署一个 Broker，ZooKeeper 使用的是单独的集群，然后创建一个 Topic，启动模拟的生产者和消费者脚本，在生产者端向 Topic 里写数据，在消费者端观察读取到的数据。

27.3 实验原理

27.3.1 Kafka 简介

Kafka 是一种高吞吐量的分布式发布订阅消息系统，它可以处理消费者规模的网站中的所有动作流数据。它提供了类似于 JMS 的特性，但是在设计实现上完全不同，此外它并不是 JMS 规范的实现。Kafka 对消息保存时根据 Topic 进行归类，发送消息者成为 Producer，消息接受者成为 Consumer，此外 Kafka 集群有多个 Kafka 实例组成，每个实例（server）成为 Broker。无论是 Kafka 集群，还是 Producer 和 Consumer 都依赖于 ZooKeeper 来保证系统可用性集群保存一些 meta 信息。如图 27-1 所示。

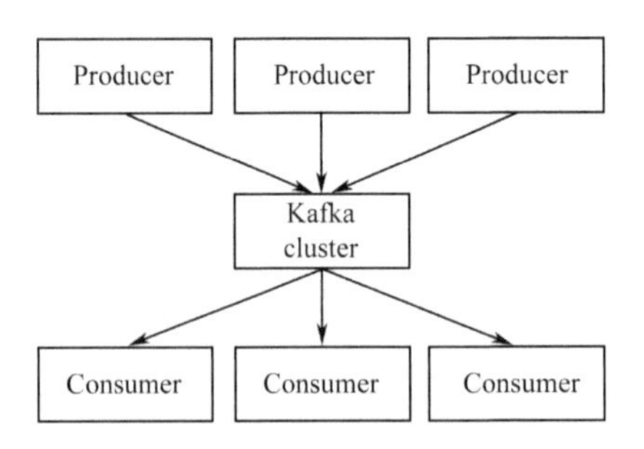

图 27-1 Kafka 简介

一个 Topic 的多个 partitions，被分布在 Kafka 集群中的多个 server 上；每个 server（Kafka 实例）负责 partitions 中消息的读写操作；此外 Kafka 还可以配置 partitions 需要备份的个数（replicas），每个 partition 将会被备份到多台机器上，以提高可用性。

基于 replicated 方案，那么就意味着需要对多个备份进行调度；每个 partition 都有一个 server 为"leader"；leader 负责所有的读写操作，如果 leader 失效，那么将会有其他 follower 来接管（成为新的 leader）；follower 只是单调的和 leader 跟进，同步消息即可。由此可见作为 leader 的 server 承载了全部的请求压力，因此从集群的整体考虑，有多少个 partitions 就意味着有多少个"leader"，Kafka 会将"leader"均衡地分散在每个实例上，来确保整体的性能稳定。

生产者：Producer 将消息发布到指定的 Topic 中，同时 Producer 也能决定将此消息归属于哪个 partition；比如基于"round-robin"方式或者通过其他的一些算法等。

消费者：本质上 Kafka 只支持 Topic。每个 Consumer 属于一个 Consumer group；反过来说，每个 group 中可以有多个 Consumer。发送到 Topic 的消息，只会被订阅此 Topic 的每个 group 中的一个 Consumer 消费。

如果所有的 Consumer 都具有相同的 group，这种情况和 queue 模式很像；消息将会在 Consumers 之间负载均衡。

如果所有的 Consumer 都具有不同的 group，那这就是"发布—订阅"；消息将会广播给所有的消费者。

在 Kafka 中，一个 partition 中的消息只会被 group 中的一个 Consumer 消费；每个 group 中 consumer 消息消费互相独立；我们可以认为一个 group 是一个"订阅"者，一个 Topic 中的每个 partions，只会被一个"订阅者"中的一个 Consumer 消费，不过一个 Consumer 可以消费多个 partitions 中的消息。Kafka 只能保证一个 partition 中的消息被某个 Consumer 消费时，消息是顺序的。事实上，从 Topic 角度来说，消息仍不是有序的。

Kafka 的设计原理决定，对于一个 Topic，同一个 group 中不能有多于 partitions 个数的 Consumer 同时消费，否则将意味着某些 Consumer 将无法得到消息。

Guarantees

（1）发送到 partitions 中的消息将会按照它接收的顺序追加到日志中。

（2）对于消费者而言，它们消费消息的顺序和日志中消息顺序一致。

（3）如果 Topic 的"replicationfactor"为 N，那么允许 N-1 个 Kafka 实例失效。

27.3.2 Kafka 使用场景

（1）Messaging

对于一些常规的消息系统，Kafka 是个不错的选择；partitons/replication 和容错，可以使 Kafka 具有良好的扩展性和性能优势。不过到目前为止，我们应该很清楚认识到，Kafka 并没有提供 JMS 中的"事务性""消息传输担保（消息确认机制）""消息分组"等企业级特性；Kafka 只能使用作为"常规"的消息系统，在一定程度上，尚未确保消息的发送与接收绝对可靠（比如，消息重发，消息发送丢失等）。

（2）Websit activity tracking

Kafka 可以作为“网站活性跟踪”的最佳工具；可以将网页/用户操作等信息发送到 Kafka 中。并实时监控，或者离线统计分析等。

（3）Log Aggregation

Kafka 的特性决定它非常适合作为“日志收集中心”；application 可以将操作日志“批量”“异步”的发送到 Kafka 集群中，而不是保存在本地或者 DB 中；Kafka 可以批量提交消息/压缩消息等，这对 Producer 端而言，几乎感觉不到性能的开支。此时 Consumer 端可以使 Hadoop 等其他系统化的存储和分析系统。

27.4 实验步骤

本实验主要演示 Kafka 的安装及简单使用，Kafka 数据保存在 ZooKeeper 上，所以该实验主要包含以下三个步骤。

27.4.1 安装 ZooKeeper 集群

安装 ZooKeeper 集群, 这个参照前面的实验操作，这里不再赘述。

27.4.2 安装 Kafka 集群

将 Kafka 安装包解压到 slave1 的/usr/cstor 目录，将 Kafka 目录所属用户改成 root:root，并将 Kafka 目录传到其他两台机器上。

```
tar -zxvf kafka_2.10-0.9.0.1.tar.gz -c /usr/cstor
chown -R root:root /usr/cstor/kafka
scp -r /usr/cstor/kafka hadoop@slave2:/usr/cstor
scp -r /usr/cstor/kafka hadoop@slave3:/usr/cstor
```

三台机器上分别进入解压目录下，在 config 目录修改 server.properties 文件，修改内容如下。

slave1 配置：

```
#broker.id
broker.id=1
#broker.port
port=9092
#host.name
host.name=slave1
#本地日志文件位置
log.dirs=/usr/cstor/kafka/logs
#Zookeeper 地址
zookeeper.connect=slave1:2181,slave2:2181,slave3:2181
```

slave2 配置：

```
#broker.id
broker.id=2
#broker.port
```

```
port=9092
#host.name
host.name=slave2
#本地日志文件位置
log.dirs=/usr/cstor/kafka/logs
#Zookeeper 地址
zookeeper.connect=slave1:2181,slave2:2181,slave3:2181
```

slave3 配置：

```
#broker.id
broker.id=3
#broker.port
port=9092
#host.name
host.name=slave3
#本地日志文件位置
log.dirs=/usr/cstor/kafka/logs
#Zookeeper 地址
zookeeper.connect=slave1:2181,slave2:2181,slave3:2181
```

然后，启动 Kafka，并验证 Kafka 功能。

进入安装目录下的 bin 目录，三台机器上分别执行以下命令启动各自的 Kafka 服务：

```
cd /usr/cstor/kafka/bin
nohup ./kafka-server-start.sh ../config/server.properties &
```

在任意一台机器上，执行以下命令（以下三行命令不要换行，是一整行）创建 Topic：

```
./kafka-topics.sh --create
--zookeeper slave1:2181,slave2:2181,slave3:2181
--replication-factor 2 --partitions 2 --topic test
```

在任意一台机器上，执行以下命令（以下三行命令不要换行，是一整行）启动模拟 Producer：

```
./kafka-console-producer.sh
--broker-list slave1:9092,slave2:9092,slave3:9092
--topic test
```

在另一台机器上，执行以下命令（以下三行命令不要换行，是一整行）启动模拟 Consumer：

```
./kafka-console-consumer.sh
--zookeeper slave1:2181,slave2:2181,slave3:2181
--topic test --from-beginning
```

27.4.3 验证消息推送

在 Producer 端输入任意信息，然后观察 Consumer 端接收到的数据，如：

```
This is Kafka producer
Hello, Kafka
```

27.5 实验结果

启动 Producer 端和 Consumer 端后，在 Kafka Producer 端发送消息后，在 Consumer 端收到信息内容如图 27-2 所示。

```
[root@slave2 bin]# ./kafka-console-consumer.sh --zookeeper slave1:2181,slave2:2181,slave3:2181 --topic test --from-beginning
This is Kafka producer
Hello, Kafka
```

图 27-2　Kafka 接收信息

实验二十八　Pig 版 WordCount

28.1　实验目的

掌握 Pig 的安装部署，了解 Pig 和 MapReduce 之间的关系，掌握 Pig Latin 编程语言，加深对 Pig 相关概念的理解。

28.2　实验要求

1. 在集群中任一节点部署 Pig（以下示例中选主节点）；
2. 在 Pig 的 Hadoop 模式下统计任意一个 HDFS 文本中的单词出现次数，并将统计结果打印出来；
3. 用 Pig 查询统计 HDFS 文件系统中结构化的数据。

28.3　实验原理

Pig 是一种探索大规模数据集的脚本语言。MapReducer 的一个主要的缺点就是开发的周期太长了。我们要编写 mapper 和 reducer，然后对代码进行编译打出 jar 包，提交到本地的 JVM 或者是 Hadoop 的集群上，最后获取结果，这个周期是非常耗时的，即使使用 Streaming（它是 Hadoop 的一个工具，用来创建和运行一类特殊的 map/reduce 作业。所谓的特殊的 map/reduce 作业可以是可执行文件或脚本本件（python、PHP、C 等）。Streaming 使用“标准输入”和“标准输出”与我们编写的 map 和 reduce 进行数据的交换。由此可知，任何能够使用“标准输入”和“标准输出”的编程语言都可以用来编写 MapReduce 程序，能在这个过程中除去代码的编译和打包的步骤，但是这一个过程还是很耗时，Pig 的强大之处就是它只要几行 Pig Latin 代码就能处理 TB 级别的数据。Pig 提供了多个命令用于检查和处理程序中的数据结构，因此它能很好地支持我们写查询。Pig 的一个很有用的特性就是它支持在输入数据中有代表性的一个小的数据集上试运行。所以，我们在处理大的数据集前可以用那一个小的数据集检查我们的程序是不是有错误的。

Pig 为大型的数据集的处理提供了更高层次的抽象。MapReducer 能够让我们自己定义 连续执行的 map 和 reduce 函数，但是数据处理往往需要很多的 MapReducer 过程才能实现，所以将数据处理要求改写成 MapReducer 模式是很复杂的。和 MapReducer 相比，Pig 提供了更加丰富的数据结构，一般都是多值和嵌套的数据结构。Pig 还提供了一套更强大的数据交换操作，包括了 MapReducer 中被忽视的“join”操作。

Pig 被设计为可以扩展的，处理路径上的每一个部分，载入、存储、过滤、分组、连接，都是可以定制的，这些操作都可以使用用户定义函数（User-Defined Function，UDF）进行修改，这些函数作用于 Pig 的嵌套数据模型。因此，它们可以在底层与 Pig 的操作集成，UDF 的另外的一个好处是它们比 MapReducer 程序开发的库更易于重用。

但是，Pig 并不适合处理所有的“数据处理”任务。和 MapReducer 一样，它是为数据批处理而设计的，如果想执行的查询只涉及一个大型数据集的一小部分数据，Pig 的实现不是很好，因为它要扫描整个数据集或其中的很大一部分。

Pig Latin 程序是由一系列的“操作”(operation)或“变换” (transformation)组成。每个操作或变换对输入进行数据处理，然后产生输出的结果。这些操作整体上描述了一个数据流，Pig 执行的环境把数据流翻译为可执行的内部表示，并运行它。在 Pig 的内部，这些变换和操作被转换成一系列的 MapReducer，但是我们一般情况下并不知道这些转换是怎么进行的，我们的主要的精力就花在数据上，而不是执行的细节上面。

在有些情况下，Pig 的表现不如 MapReducer 程序。综上所述，要么话花大量的时间来优化 Java MapReducer 程序，要么使用 Pig Latin 来编写查询确实能节约时间。

28.4 实验步骤

本实验只要演示 Pig 的安装部署和 Pig 在 Hadoop 模式下的 WordCount 和结构化数据查询统计示例演示，只要包含以下三个步骤。

将 Pig 安装包解压到/usr/cstor 目录。

```
[root@master usr]#   tar -zxvf pig-0.15.0.tar.gz
[root@master usr]#   mv   pig-0.15.0   /usr/cstor
tar -zxvf pig-0.15.0.tar.gz -c /usr/cstor
```

进入解压目录下，修改 bin 目录下 pig 文件，添加如下内容：

```
#指定 Java 安装路径
JAVA_HOME=/usr/local/jdk1.7.0_79
#指定 Hadoop 安装路径
HADOOP_HOME=/usr/cstor/hadoop
#指定 Hadoop 配置文件的路径
PIG_CLASSPATH=/usr/cstor/hadoop/etc/hadoop
```

Pig 配置完成后，在安装 Pig 的节点进入 hadoop 安装目录下的 sbin 目录，并启动 historyserver 服务：

```
[root@master hadoop]#   cd /usr/cstor/hadoop/sbin
[root@master sbin]#   ./mr-jobhistory-daemon.sh start historyserver
```

（1）单词计数

在 HDFS 文件系统创建一个 pig 目录，并从本地上传任意一个文件到目录中，为后面的 Pig 单词统计准备数据：

```
cd /usr/cstor/hadoop/bin
hdfs dfs -mkdir /pig
```

```
hdfs dfs -copyFromLocal ../README.txt /pig
```

HDFS 数据准备好后，再进入 Pig 的 bin 目录执行下面的命令，进入 Pig 编程界面：

```
    [root@master cstor]# cd    /usr/cstor/pig/bin
    [root@master bin]# ./pig
```

代码实现如下：

```
#按行读 HDFS 文件
A = LOAD '/pig/README.txt' AS (line: chararray) ;
#将每行单词用 tab、逗号，句号以及空格分隔单词
B = foreach A generate flatten(TOKENIZE(line,'\t ,.'))as word;
#按单词分组，将相同的单词归并到一起
C = group B by word;
#统计每个单词的个数
D = foreach C generate group, COUNT(B) as count;
#将结果显示出来
DUMP D;
```

（2）结构化数据查询统计

在 Pig 节点的/root/data/28 目录下新建文本文件 pigdata.txt：

```
[root@master 28]# touch pigdata.txt
```

内容如下：

zhangsan,m,28

lisi,f,26

wangwu,m,20

将 pigdata.txt 上传到 HDFS 的/pig 目录下：

```
cd /usr/cstor/hadoop/bin
[root@master bin]#    hdfs dfs -copyFromLocal /root/data/28/pigdata.txt /pig
```

进入 Pig 编程界面：

```
cd /usr/cstor/pig/bin
[root@master bin]# ./pig
```

代码实现如下：

```
#按行读 HDFS 文件，并将数据按逗号分隔解析成姓名，性别，年龄
A = LOAD '/pig/pigdata.txt' USING PigStorage(',') AS(name:chararray, sex:chararray, age:int);
#打印 A 中数据
DUMP A;
#只查询 name 字段，并打印
B = FOREACH A GENERATE name;
DUMP B;
#查询所有人员数据，并将年龄加一
C = FOREACH A GENERATE name, sex, age+1;
DUMP C;
#查询年龄大于 20 的人员
D = FILTER A BY (age > 20);
DUMP D;
```

```
#统计数据总条数
E = GROUP A ALL ;
F = FOREACH E GENERATE COUNT (A);
DUMP F;
```

28.5 实验结果

Pig 客户端输出结果如图 28-1 所示。

```
Success!

Job Stats (time in seconds):
JobId   Maps    Reduces MaxMapTime      MinMapTime      AvgMapTime      MedianMapTime   MaxReduceTime   MinReduceTime   AvgReduceTime   MedianRed
ucetime Alias   Feature Outputs
job_1478178184207_0002  1       1       3       3       3       3       3       3       3       3       A,B,C,D GROUP_BY,COMBINER       hdfs://ma
ster:9000/tmp/temp-1058269208/tmp-2034324266,

Input(s):
Successfully read 31 records (1726 bytes) from: "/pig/README.txt"

Output(s):
Successfully stored 136 records (1950 bytes) in: "hdfs://master:9000/tmp/temp-1058269208/tmp-2034324266"

Counters:
Total records written : 136
Total bytes written : 1950
Spillable Memory Manager spill count : 0
Total bags proactively spilled: 0
Total records proactively spilled: 0

Job DAG:
job_1478178184207_0002
```

图 28-1 Pig 客户端输出结果

Pig 客户端单词统计结果如图 28-2 所示。

```
(1,1)
(C,1)
(S,1)
(U,1)
(as,1)
(by,1)
(if,1)
(in,1)
(is,1)
(it,1)
(of,5)
(on,2)
(or,2)
(to,2)
(13),1)
(740,1)
(BIS,1)
(ENC,1)
(For,1)
(SSL,1)
```

图 28-2 单词统计

查询全部人员结果如图 28-3 所示。

```
(zhangsan,m,28)
(lisi,f,26)
(wangwu,m,20)
grunt>
```

图 28-3 查询全部人员

查询全部人员姓名结果如图 28-4 所示。

```
(zhangsan)
(lisi)
(wangwu)
grunt>
```

图 28-4　查询全部人员姓名

查询所有人信息，并且年龄加一，结果如图 28-5 所示。

```
(zhangsan,m,29)
(lisi,f,27)
(wangwu,m,21)
grunt>
```

图 28-5　查询年龄加一

查询年龄大于 20 的人员，结果如图 28-6 所示。

```
(zhangsan,m,28)
(lisi,f,26)
grunt>
```

图 28-6　查询年龄大于 20 的人员

统计数据总条数，结果如图 28-7 所示。

```
(3)
grunt>
```

图 28-7　统计数据总条数

实验二十九　Redis 部署与简单使用

29.1　实验目的

1. 熟悉 CentOS 系统；
2. 学习从 C++源代码编译成可执行文件并安装运行；
3. 学习安装 Redis；
4. 学习配置 Redis；
5. 学会简单使用 Redis。

29.2　实验要求

在 client 机上，安装配置并简单使用 Redis。

29.3　实验原理

29.3.1　CentOS 简介

社区企业操作系统（Community Enterprise Operating System，CentOS）是 Linux 发行版之一。它是来自 Red Hat Enterprise Linux 依照开放源代码规定释出的源代码所编译而成。由于出自同样的源代码，因此有些要求高度稳定性的服务器以 CentOS 替代商业版的 Red Hat Enterprise Linux（RHEL）使用。两者的不同在于 CentOS 并不包含封闭源代码软件，最新的 CentOS 版本是 7.2，内核版本是 3.10.0。

29.3.2　CentOS 与 RHEL 关系

CentOS 与 RHEL 的关系：RHEL 在发行的时候，有两种方式。一种是二进制的发行方式，另一种是源代码的发行方式。无论是哪一种发行方式，你都可以免费获得（如从网上下载），并再次发布。但如果你使用了他们的在线升级，包括补丁或咨询服务，就必须要付费。RHEL 一直都提供源代码的发行方式，CentOS 就是将 RHEL 发行的源代码重新编译一次，形成一个可使用的二进制版本。由于 Linux 的源代码是 GNU，所以从获得 RHEL 的源代码到编译成新的二进制，都是合法。只是 red hat 是商标，所以必须在新的发行版里将 red hat 的商标去掉。red hat 对这种发行版的态度是：“我们其实并不反对这种发行版，真正向我们付费的用户，他们重视的并不是系统本身，而是我们所提供的商业服务。”所以，CentOS 可以得到 RHEL 的所有功能，甚至是更好的软件。但 CentOS 并不向用户提供商业支持，当然也不负任何商业责任。如果你要将你的 RHEL 转到 CentOS 上，因为你不希望为 RHEL 升级而付费。当然，你必须有丰富 Linux 使用经验，因此 RHEL 的商业技术支持对你来说并不重要。但如果你是单纯的业务型企业，那么还是建议你选购 RHEL 软件并购买相应服务。这样可以节省你的 IT 管理费用，并可得到专业服务。一句话，选用 CentOS 还是 RHEL，取决于你所在公司是否拥有相应的技术力量。

29.3.3　make 简介

Linux 下 make 命令是系统管理员和程序员用的最频繁的命令之一。管理员用它通过命令行来编译和安装很多开源的工具，程序员用它来管理他们大型复杂的项目编译问题。make 命令通常和 Makefile 一起使用。

在开发一个系统时，一般是将一个系统分成几个模块，这样做提高了系统的可维护性，但由于各个模块间不可避免存在关联，所以当一个模块改动后，其他模块也许会有所更新，当然对小系统来说，手工编译连接是没问题，但是如果是一个大系统，存在很多个模块，那么手工编译的方法就不适用了。为此，在 Linux 系统中，专门提供了一个 make 命令来自动维护目标文件，与手工编译和连接相比，make 命令的优点在于他只更新修改过的文件（在 Linux 中，一个文件被创建或更新后有一个最后修改时间，make 命令就是通过这个最后修改时间来判断此文件是否被修改），而对没修改的文件则置之不理，并且 make 命令不会漏掉一个需要更新的文件。

文件和文件间或模块或模块间有可能存在依赖关系，make 命令也是依据这种依赖关系来进行维护的，所以我们有必要了解什么是依赖关系；make 命令当然不会自己知道这些依赖关系，而需要程序员将这些依赖关系写入一个叫 Makefile 的文件中。Makefile 文件中包含着一些目标，通常目标就是文件名，对每一个目标，提供了实现这个目标的一组命令以及和这个目标有依赖关系的其他目标或文件名，以下是一个简单的 Makefile 的例子。

```
#一个简单的 Makefile
prog:prog1.o prog2.o
	gcc prog1.o prog2.o -o prog
prog1.o:prog1.c lib.h
	gcc -c -I. -o prog1.o prog1.c
prog2.o:prog2.c
	gcc -c prog2.c
```

以上 Mamefile 中定义了三个目标：prog、prog1 和 prog2，冒号后是依赖文件列表。对于第一个目标文件 prog 来说，它有两个依赖文件：prog1.o 和 prog2.o，任何一个依赖文件更新，prog 也要随之更新，命令 gcc prog1.o prog2.o -o prog 是生成 prog 的命令。make 检查目标是否需要更新时采用递归的方法，递归从底层向上对过时目标进行更新，只有当一个目标所依赖的所有目标都为最新时，这个目标才会被更新。以上面的

Makefile 为例，我们修改了 prog2.c，执行 make 时，由于目标 prog 依赖 prog1.o 和 prog2.o，所以要先检查 prog1.o 和 prog2.o 是否过时，目标 prog1.o 依赖 prog1.c 和 lib.h，由于我们并没修改这两个文件，所以他们都没有过期，接下来再检查目标 prog2.o，他依赖 prog2.c，由于我们修改了 prog2.c，所以 prog2.c 比目标文件 prog2.o 要新，即 prog2.o 过期，而导致了依赖 prog2.o 的所有目标都过时；这样 make 会先更新 prog2.o 再更新 prog。

29.3.4 Redis 简介

Redis 是一个开源的使用 ANSI C 语言编写、支持网络、可基于内存亦可持久化的日志型、Key-Value 数据库，并提供多种语言的 API。从 2010 年 3 月 15 日起，Redis 的开发工作由 VMware 主持。从 2013 年 5 月开始，Redis 的开发由 Pivotal 赞助。

Redis 是一个 Key-Value 存储系统。和 Memcached 类似，它支持存储的 Value 类型相对更多，包括 string（字符串）、list（链表）、set（集合）、zset（sorted set——有序集合）和 hash（哈希类型）。这些数据类型都支持 push/pop、add/remove 及取交集并集和差集及更丰富的操作，而且这些操作都是原子性的。在此基础上，Redis 支持各种不同方式的排序。与 Memcached 一样，为了保证效率，数据都是缓存在内存中。区别是 Redis 会周期性地把更新的数据写入磁盘或者把修改操作写入追加的记录文件，并且在此基础上实现了 master-slave（主从）同步。

Redis 是一个高性能的 Key-Value 数据库，Redis 的出现，很大程度补偿了 Memcached 这类 Key/Value 存储的不足，在部分场合可以对关系数据库起到很好的补充作用。它提供了 Java，C/C++，C#，PHP，JavaScript，Perl，Object-C，Python，Ruby，Erlang 等客户端，使用很方便。

Redis 支持主从同步。数据可以从主服务器向任意数量的从服务器上同步，从服务器可以是关联其他从服务器的主服务器。这使得 Redis 可执行单层树复制。存盘可以有意无意地对数据进行写操作。由于完全实现了发布/订阅机制，使得从数据库在任何地方同步树时，可订阅一个频道并接收主服务器完整的消息发布记录。同步对读取操作的可扩展性和数据冗余很有帮助。

29.4 实验步骤

本实验包括安装、配置与使用三部分，下面按此三部分依次讲述。

29.4.1 安装配置启动

首先，登录 client 机，在该机上，依次输入下述命令安装即可，注意“make”与“make install”命令执行时会有大量输出（故此处并未截图），请读者确保这两个命令都在“/usr/cstor/redis”目录下执行即可，其他暂不必理会。

```
[root@client ~]# cd /usr/cstor/redis/
[root@client redis]# make
```

```
[root@client redis]# make  install
```

其次，读者使用“vim”，编辑“/usr/cstor/redis/redis.conf”文件，定位到下述内容所在行：

```
bind  127.0.0.1
```

将上述的“127.0.0.1”换成本机 IP，笔者此处为“10.1.1.76”。

```
bind  172.17.0.64
```

最后，使用“redis-server”命令，指定上述配置文件，启动 Redis，操作如图 29-1 所示。

```
[root@client ~]#  redis-server  /usr/cstor/redis/redis.conf  &
```

```
[root@client ~]#
[root@client ~]# redis-server  /usr/cstor/redis/redis.conf  &
[1] 3173
[root@client ~]#
```

图 29-1　以后台服务方式启动 Redis 数据库

29.4.2　使用 Redis

连接本地 Redis

```
[root@client redis]$ src/redis-cli
```

连接远程 Redis

```
[root@master redis]$ src/redis-cli  -h  client
```

向 Redis 里写入数据

```
[root@client redis]$ src/redis-cli -h 172.17.0.15
172.17.0.15:6379> set chengshi sh,bj,sz,nj,hf
172.17.0.15:6379> get  chengshi
```

29.5　实验结果

请参考 29.4.2 的实验过程。

实验三十　MapReduce 与 Spark 读写 Redis

30.1　实验目的

1. 使用 MapReduce 访问 Redis 数据；
2. 使用 Spark 访问 Redis 数据。

30.2　实验要求

1. 在 client 机上，使用 MapReduce 代码读取 Redis 数据；
2. 在 client 机上，使用 Spark 代码读取 Redis 数据。

30.3　实验原理

假定现有一个大为 1000GB 的大表 big.txt 和一个大小为 10GB 的小表 small.txt，请基于 MapReduce 思想编程实现判断小表中单词在大表中出现次数。也即所谓的"扫描大表、加载小表"。

为解决上述问题，可开启 10 个 Map。这样，每个 Map 只需处理总量的 1/10，将大大加快处理。而在单独 Map 内，由于 10GB 的 small.txt 依旧非常巨大，显然不适宜 HashSet 加载到内存等措施，此时可借助第三方存储介质（如 Redis），在 Map 阶段先加载部分数据，然后再加载部分数据，甚至可直接将结果写入 Redis，如图 30-1 所示。

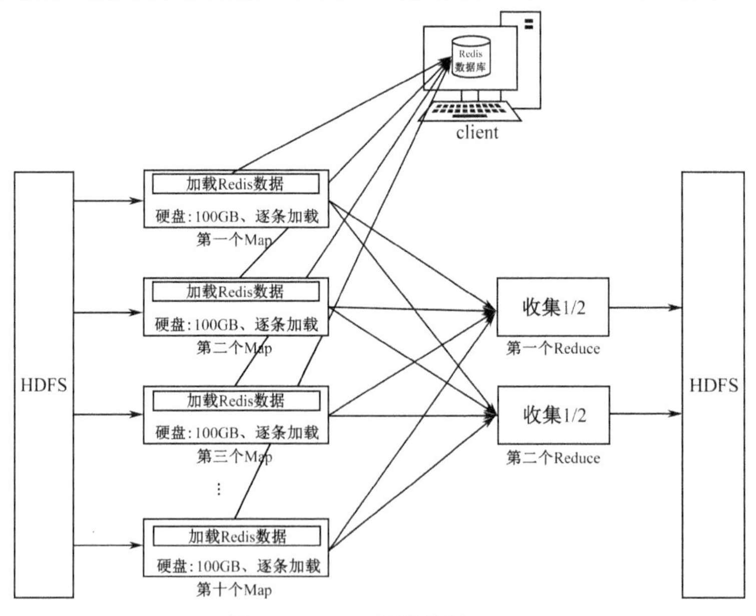

图 30-1　Map 阶段访问 Redis

30.4 实验步骤

主要包括 MapReduce 读取 Redis 代码和 Spark 读取 Redis 数据代码。

30.4.1 MapReduce 读取 Redis

首先是准备数据阶段，准备好 big.txt 并将其上传至 HDFS，准备好待查城市名并将其导入 Redis。接着为编程阶段，编写 MapReduce 程序，在此程序 Map 阶段，取出 Redis 里待查城市，顺序扫描数据块里数据，是该城市则输出，不是则不做任何操作。最后，在集群上执行该程序。

1. 准备 HDFS 数据

首先，登录 client 机或者 master 机器，确认该机上存在"/root/data/30/big.txt"（如果不存在请自己手动添加这样的文本数据），如图 30-2 所示。

```
[root@client ~]# cat /root/data/30/big.txt
aaa bbb ccc nanjing eee fff ggg
hhh iii jjj nanjing lll mmm nnn
000 111 222 333 nanjing 555 666 777 888 999
ooo ppp nanjing rrr sss ttt
uuu vvv www nanjing yyy zzz
[root@client ~]#
```

图 30-2 确认本地文件 big.txt

其次，登录到 client 机或者 master 机器上，查看 HDFS 里是否已存在目录"/user/root/redis/in"，若不存在，使用下述命令新建该目录。

```
[root@client ~]# /usr/cstor/hadoop/bin/hdfs dfs -mkdir -p /user/root/redis/in
```

再次，使用下述命令将 client 机本地文件"/root/data/30/big.txt"上传至 HDFS 的"/user/root/redis/in"目录：

```
[root@client ~]# /usr/cstor/hadoop/bin/hdfs dfs -put /root/data/30/big.txt /user/root/redis/in
```

最后，请进入 HDFS Web 页面，确认 HDFS 上文件与内容，如图 30-3 所示。

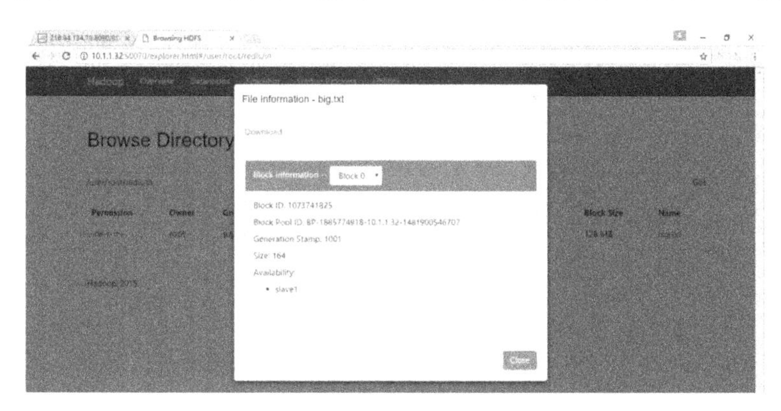

图 30-3 确认 HDFS 上 big.txt 文件位置与内容

也可以使用命令行查看文件：

```
[root@client ~]# /usr/cstor/hadoop/bin/hdfs   dfs   -ls   /user/root/redis/in
```

2. 准备 Redis 数据

首先，参考下述命令登录到 Redis 数据库，进入 Redis 的 src 目录：cd /usr/cstor/Redis/src，然后登录 redis。

```
[root@client ~]# redis-cli   -h 10.1.1.36（这里是运行 redis 服务器的主机的 ip）
```

接着，向 Redis 数据库里写入<city,nanjing>，参考命令如下：

```
10.1.1.36:6379> set   city   nanjing
10.1.1.36:6379> get   city
```

3. 编程 MapReduce 程序

首先，打开 Eclipse，依次点击“File→New→Other…→Map/Reduce Project”，在弹出的“New MapReduce Project Wizard”对话框中 “Project name:”一栏填写项目名“SharedLargeMemory”，然后直接单击该对话框的“Finish”按钮。最后，新建 LargeMemory 类并指定包名（代码中为 cn.cstor.redis），在 LargeMemory.java 文件中，依次写入如下代码：

```
package cn.cstor.redis;
import java.io.IOException;
import org.apache.hadoop.conf.Configuration;
import org.apache.hadoop.fs.Path;
import org.apache.hadoop.io.IntWritable;
import org.apache.hadoop.io.Text;
import org.apache.hadoop.mapreduce.Job;
import org.apache.hadoop.mapreduce.Mapper;
import org.apache.hadoop.mapreduce.Reducer;
import org.apache.hadoop.mapreduce.lib.input.FileInputFormat;
import org.apache.hadoop.mapreduce.lib.output.FileOutputFormat;
import redis.clients.jedis.Jedis;

public class LargeMemory {
public static class TokenizerMapper extends Mapper<Object, Text, Text, IntWritable> {
private final static IntWritable one = new IntWritable(1);
Jedis jedis = null;
protected void setup(Context context) throws IOException, InterruptedException {
jedis = new Jedis(context.getConfiguration().get("redisIP"));
System.out.println("setup ok *^_^* ");
}
public void map(Object key, Text value, Context context) throws IOException, InterruptedException {
String[] values = value.toString().split(" ");
for (int i = 0; i < values.length; i++) {
```

```
if (jedis.get("city").equals(values[i])) {
context.write(new Text(values[i]), one);
}
}
}
}
public static class IntSumReducer extends Reducer<Text, IntWritable, Text, IntWritable> {
private IntWritable result = new IntWritable();
public void reduce(Text key, Iterable<IntWritable> values, Context context)
throws IOException, InterruptedException {
int sum = 0;
for (IntWritable val : values) {
sum += val.get();
}
result.set(sum);
context.write(key, result);
}
}
public static void main(String[] args) throws Exception {
Configuration conf = new Configuration();
conf.set("redisIP", args[0]);
Job job = Job.getInstance(conf, "RedisDemo");
job.setJarByClass(LargeMemory.class);
job.setMapperClass(TokenizerMapper.class);
job.setReducerClass(IntSumReducer.class);
job.setMapOutputKeyClass(Text.class);
job.setMapOutputValueClass(IntWritable.class);
job.setOutputKeyClass(Text.class);
job.setOutputValueClass(IntWritable.class);
FileInputFormat.addInputPath(job, new Path(args[1]));
FileOutputFormat.setOutputPath(job, new Path(args[2]));
System.exit(job.waitForCompletion(true) ? 0 : 1);
}
}
```

由于该程序依赖 Redis，因此还需添加 Redis 客户端 jar 包。实际操作时，选中该项目，“New→Folder”，在弹出的对话框中填写“lib”，接着将“jedis-2.1.0.jar”（这个 jedis-2.1.0.jar 包如果实验环境下找不到可以从网络上下载也是同样可以使用的）复制到该文件夹（“lib”）下，最后，选中“jedis-2.1.0.jar”，依次点击“Build Path→Add to Build Path”完成添加依赖包，如图 30-4 所示。

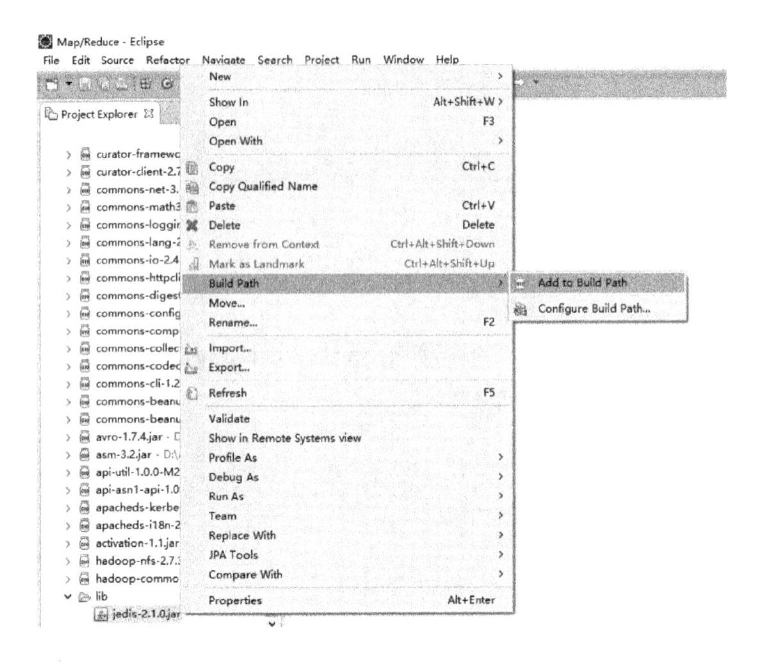

图 30-4　将 jar 包添加到 Build Path

至此，已完成代码开发，图 30-5 为本项目结构图，请读者对照该图，分析项目结构图中各模块。

```
package cn.cstor.redis;

import java.io.IOException;

public class LargeMemory {
    public static class TokenizerMapper extends Mapper<Object, Text, Text, IntWritable> {
        private final static IntWritable one = new IntWritable(1);
        Jedis jedis = null;
        protected void setup(Context context) throws IOException, InterruptedException {
            jedis = new Jedis(context.getConfiguration().get("redisIP"));
            System.out.println("setup ok *^_^* ");
        }
        public void map(Object key, Text value, Context context) throws IOException, InterruptedExcep
            String[] values = value.toString().split(" ");
            for (int i = 0; i < values.length; i++) {
                if (jedis.get("city").equals(values[i])) {
                    context.write(new Text(values[i]), one);
                }
            }
        }
    }
    public static class IntSumReducer extends Reducer<Text, IntWritable, Text, IntWritable> {
        private IntWritable result = new IntWritable();
        public void reduce(Text key, Iterable<IntWritable> values, Context context)
                throws IOException, InterruptedException {
            int sum = 0;
            for (IntWritable val : values) {
                sum += val.get();
            }
            result.set(sum);
            context.write(key, result);
        }
```

图 30-5　本项目集成开发界面

4. 打包该程序

待代码编写结束，选中该项目，依次点击“Export→Java→JAR file”，弹出对话框，在图 30-6 中填写打包位置，接着 Finish 即可。笔者此处打包时包名及其位置为“C:\Users\allen\Desktop\SharedLargeMemory.jar”。

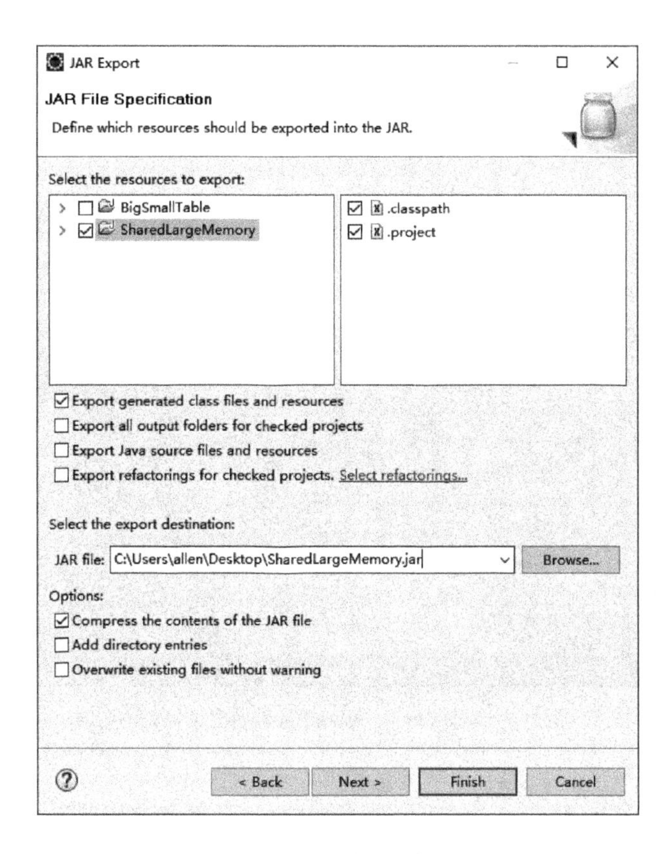

图 30-6 项目打包

5. 执行代码

首先，使用xftp或者一体机提供的工具将"C:\Users\allen\Desktop\SharedLargeMemory.jar"上传至client机，此处上传至"/root/SharedLargeMemory.jar"。

接着，登录client机上，使用下述命令提交SharedLargeMemory.jar任务。

```
[root@client ~]# /usr/cstor/hadoop/bin/hadoop jar /root/SharedLargeMemory.jar
cn.cstor.redis.LargeMemory 10.1.1.36 /user/root/redis/in/big.txt /user/root/redis/SLMResult
```

30.4.2 Spark读取Redis

和MapReduce相比，使用Spark读写Redis则简单的多，首先，登录Redis准备数据；其次，在启动Spark时指定Redis包；最后，在Spark交互式执行界面中直接编写代码访问Redis即可。具体操作如下。

1. 准备Redis数据

登录Redis，向Redis数据库添加数据。

```
[root@client ~]$ redis-cli -h client
172.17.0.15:6379> SET chengshi sh,bj,sz,nj,hf
172.17.0.15:6379> get chengshi
```

2. 启动 Spark Shell

首先启动 Spark 集群，再启动 Spark Shell，由于 Spark 访问 Redis 时需要 Redis 客户端 jedis，故此处启动 Spark 时需指定 jedis 包（如果 Redis 的 jar 包不存在，可以手动放到/usr/cstor/redis/这个目录下），启动命令如下：

```
[root@client ~]$/usr/cstor/spark/bin/spark-shell --master spark://master:7077 --jars /usr/cstor/redis/jedis-2.1.0.jar
```

启动成功后界面如图 30-7 所示。

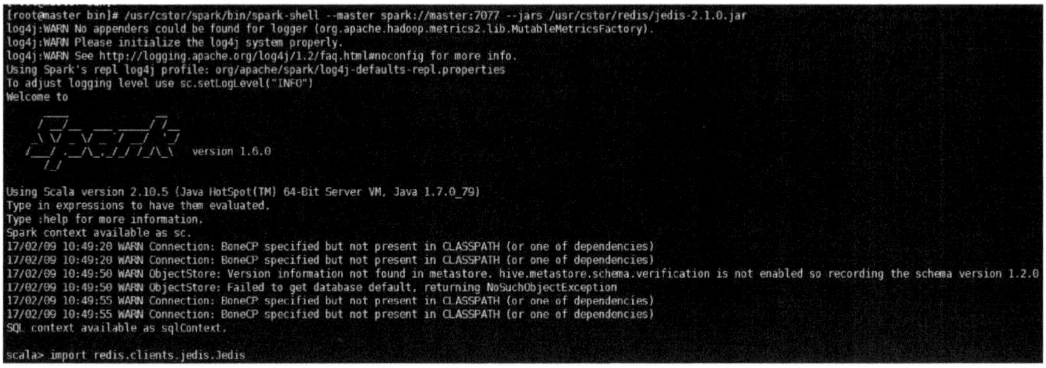

```
[root@master bin]# /usr/cstor/spark/bin/spark-shell --master spark://master:7077 --jars /usr/cstor/redis/jedis-2.1.0.jar
log4j:WARN No appenders could be found for logger (org.apache.hadoop.metrics2.lib.MutableMetricsFactory).
log4j:WARN Please initialize the log4j system properly.
log4j:WARN See http://logging.apache.org/log4j/1.2/faq.html#noconfig for more info.
Using Spark's repl log4j profile: org/apache/spark/log4j-defaults-repl.properties
To adjust logging level use sc.setLogLevel("INFO")
Welcome to
                                  version 1.6.0

Using Scala version 2.10.5 (Java HotSpot(TM) 64-Bit Server VM, Java 1.7.0_79)
Type in expressions to have them evaluated.
Type :help for more information.
Spark context available as sc.
17/02/09 10:49:20 WARN Connection: BoneCP specified but not present in CLASSPATH (or one of dependencies)
17/02/09 10:49:20 WARN Connection: BoneCP specified but not present in CLASSPATH (or one of dependencies)
17/02/09 10:49:50 WARN ObjectStore: Version information not found in metastore. hive.metastore.schema.verification is not enabled so recording the schema version 1.2.0
17/02/09 10:49:50 WARN ObjectStore: Failed to get database default, returning NoSuchObjectException
17/02/09 10:49:55 WARN Connection: BoneCP specified but not present in CLASSPATH (or one of dependencies)
17/02/09 10:49:55 WARN Connection: BoneCP specified but not present in CLASSPATH (or one of dependencies)
SQL context available as sqlContext.

scala> import redis.clients.jedis.Jedis
```

图 30-7　Spark Shell 界面

3. 编写访问代码

在交互代码中连接 Redis（注意 Redis 的 IP 地址修改成自己的），读者可按下述命令操作：

```
scala> import redis.clients.jedis.Jedis
import redis.clients.jedis.Jedis
scala> var jd=new Jedis("172.17.0.15",6379)
jd: redis.clients.jedis.Jedis = redis.clients.jedis.Jedis@7dae9ff4
scala> var str=jd.get("chengshi")
str: String = sh,bj,sz,nj,hf
scala> var strList=str.split(",")
strList: Array[String] = Array(sh, bj, sz, nj, hf)
scala> val a = sc.parallelize(strList, 3)
a: org.apache.spark.rdd.RDD[String] = ParallelCollectionRDD[0] at parallelize at <console>:28
scala> val b = a.keyBy(_.length)
b: org.apache.spark.rdd.RDD[(Int, String)] = MapPartitionsRDD[1] at keyBy at <console>:30
scala> b.collect
res0: Array[(Int, String)] = Array((2,sh), (2,bj), (2,sz), (2,nj), (2,hf))
```

30.5 实验结果

30.5.1 MapReduce 读取 Redis 实验

实验结果如图 30-8 所示。

```
17/02/08 17:40:31 INFO mapreduce.Job: Counters: 35
	File System Counters
		FILE: Number of bytes read=274294
		FILE: Number of bytes written=832836
		FILE: Number of read operations=0
		FILE: Number of large read operations=0
		FILE: Number of write operations=0
		HDFS: Number of bytes read=328
		HDFS: Number of bytes written=10
		HDFS: Number of read operations=13
		HDFS: Number of large read operations=0
		HDFS: Number of write operations=4
	Map-Reduce Framework
		Map input records=5
		Map output records=5
		Map output bytes=60
		Map output materialized bytes=76
		Input split bytes=110
		Combine input records=0
		Combine output records=0
		Reduce input groups=1
		Reduce shuffle bytes=76
		Reduce input records=5
		Reduce output records=1
		Spilled Records=10
		Shuffled Maps =1
		Failed Shuffles=0
		Merged Map outputs=1
		GC time elapsed (ms)=0
		Total committed heap usage (bytes)=1029701632
	Shuffle Errors
		BAD_ID=0
		CONNECTION=0
		IO_ERROR=0
		WRONG_LENGTH=0
		WRONG_MAP=0
		WRONG_REDUCE=0
	File Input Format Counters
		Bytes Read=164
	File Output Format Counters
		Bytes Written=10
```

图 30-8 实验结果（一）

30.5.2 Spark 读取 Redis 实验

实验结果如图 30-9 所示。

```
scala> import redis.clients.jedis.Jedis
import redis.clients.jedis.Jedis

scala> var jd=new Jedis("10.1.14.85",6379);
jd: redis.clients.jedis.Jedis = redis.clients.jedis.Jedis@9543149

scala> var str=jd.get("chengshi")
str: String = sh,bj,sz,nj,hf

scala> var strList=str.split(",")
strList: Array[String] = Array(sh, bj, sz, nj, hf)

scala> val a = sc.parallelize(strList, 3)
a: org.apache.spark.rdd.RDD[String] = ParallelCollectionRDD[0] at parallelize at <console>:38

scala> val b = a.keyBy(_.length)
b: org.apache.spark.rdd.RDD[(Int, String)] = MapPartitionsRDD[1] at keyBy at <console>:40

scala>  b.collect
res0: Array[(Int, String)] = Array((2,sh), (2,bj), (2,sz), (2,nj), (2,hf))
```

图 30-9 实验结果（二）

实验三十一　MongoDB 实验：读写 MongoDB

31.1　实验目的

1. 了解 NoSQL 数据库的原理；
2. 理解 NoSQL 数据库的结构；
3. 比较 MongoDB 和 Hbase 的区别；
4. 对 MongoDB 的存储格式有一定了解；
5. 能正确使用 MongoDB 并能进行简单使用。

31.2　实验要求

1. 正确地搭建 MongoDB 数据库环境；
2. 能正常启动 MongoDB 的服务和服务的连接；
3. 能在 MongoDB 的 Shell 中进行一些简单的使用。

31.3　实验原理

MongoDB 是一个基于分布式文件存储的数据库，由 C++语言编写，旨在为 Web 应用提供可扩展的高性能数据存储解决方案。

特点：高性能、易部署、易使用，存储数据非常方便。

主要功能特性有：

（1）面向集合存储，易存储对象类型的数据；

（2）模式自由；

（3）支持动态查询；

（4）支持完全索引，包含内部对象；

（5）支持查询；

（6）支持复制和故障恢复；

（7）使用高效的二进制数据存储，包括大型对象（如视频等）；

（8）自动处理碎片，以支持云计算层次的扩展性；

（9）支持 Ruby，Python，Java，C++，PHP 等多种语言；

（10）文件存储格式为 BSON（一种 JSON 的扩展）；

（11）可通过网络访问。

所谓“面向集合”（Collenction-Oriented），意思是数据被分组存储在数据集中，被称为一个集合（Collenction）。每个集合在数据库中都有一个唯一的标识名，并且可以包含无限数目的文档。集合的概念类似关系型数据库（RDBMS）里的表（Table），不同的是它不需要定义任何模式（schema）。

模式自由（schema-free），意味着对于存储在 mongodb 数据库中的文件，我们不需要知道它的任何结构定义。如果需要的话，你完全可以把不同结构的文件存储在同一个数据库里。

存储在集合中的文档，被存储为键-值对的形式。键用于唯一标识一个文档，为字符串类型，而值则可以是各种复杂的文件类型。我们称这种存储形式为 BSON（Binary JSON）。

31.4 实验步骤

本实验总体分为几个步骤：

（1）启动 MongoDB 服务；

（2）连接 MongoDB 服务；

（3）启动 MongoDB 的 Shell，执行一些简单命令。

详细如下。

31.4.1 启动 MongoDB

首先在 MongoDB 的安装目录下建立一个数据目录，然后进入 MongoDB 安装目录下的 bin 目录，启动 MongoDB 服务。

```
[root@master ~]# cd /usr/cstor/mongodb/
[root@master mongodb]# mkdir data
[root@master mongodb]# bin/mongod --dbpath ./data &
```

31.4.2 连接使用 MongoDB

使用命令连接到 MongoDB 服务。

```
[root@master mongodb]# bin/mongo
```

31.4.3 连接启动 MongoDB 的 Shell，执行一些简单命令

连接到 MongoDB 之后，进入 Shell 环境之后执行如下的简单操作，以此来熟悉 MongoDB 的操作。

1. 创建一个 Collection

创建一个 Collection 如图 31-1 所示。

```
> db.createCollection("weather");
```

```
1 master    2 master    +
> db.createCollection("weather");
{ "ok" : 1 }
>
```

图 31-1　创建 Collection

2. 插入单条数据

插入单条数据如图 31-2 所示。

```
> db.weather.save({temp:31,location:"nanjing"});
```

```
> db.weather.save({temp:31,location:"nanjing"});
WriteResult({ "nInserted" : 1 })
>
```

图 31-2　插入单条数据

3. 插入多条数据

插入多条数据如图 31-3 所示。

```
> for(var i = 0;i < 30;i++) { db.weather.save({tep:i,location:"nanjing"}) };
```

```
1 master    2 master    +
>  for(var i = 0;i < 30;i++) { db.weather.save({tep:i,location:"nanjing"}) };
WriteResult({ "nInserted" : 1 })
>
```

图 31-3　插入多条数据

4. 简单遍历

简单遍历如图 31-4 所示。

```
> db.weather.find();
```

```
1 master    2 master    +
> db.weather.find();
{ "_id" : ObjectId("5821d101bf9520f3f8bfd1d6"), "temp" : 31, "location" : "nanjing" }
{ "_id" : ObjectId("5821d12dbf9520f3f8bfd1d7"), "tep" : 0, "location" : "nanjing" }
{ "_id" : ObjectId("5821d12dbf9520f3f8bfd1d8"), "tep" : 1, "location" : "nanjing" }
{ "_id" : ObjectId("5821d12dbf9520f3f8bfd1d9"), "tep" : 2, "location" : "nanjing" }
{ "_id" : ObjectId("5821d12dbf9520f3f8bfd1da"), "tep" : 3, "location" : "nanjing" }
{ "_id" : ObjectId("5821d12dbf9520f3f8bfd1db"), "tep" : 4, "location" : "nanjing" }
{ "_id" : ObjectId("5821d12dbf9520f3f8bfd1dc"), "tep" : 5, "location" : "nanjing" }
{ "_id" : ObjectId("5821d12dbf9520f3f8bfd1dd"), "tep" : 6, "location" : "nanjing" }
{ "_id" : ObjectId("5821d12dbf9520f3f8bfd1de"), "tep" : 7, "location" : "nanjing" }
{ "_id" : ObjectId("5821d12dbf9520f3f8bfd1df"), "tep" : 8, "location" : "nanjing" }
{ "_id" : ObjectId("5821d12dbf9520f3f8bfd1e0"), "tep" : 9, "location" : "nanjing" }
{ "_id" : ObjectId("5821d12dbf9520f3f8bfd1e1"), "tep" : 10, "location" : "nanjing" }
{ "_id" : ObjectId("5821d12dbf9520f3f8bfd1e2"), "tep" : 11, "location" : "nanjing" }
{ "_id" : ObjectId("5821d12dbf9520f3f8bfd1e3"), "tep" : 12, "location" : "nanjing" }
{ "_id" : ObjectId("5821d12dbf9520f3f8bfd1e4"), "tep" : 13, "location" : "nanjing" }
{ "_id" : ObjectId("5821d12dbf9520f3f8bfd1e5"), "tep" : 14, "location" : "nanjing" }
{ "_id" : ObjectId("5821d12dbf9520f3f8bfd1e6"), "tep" : 15, "location" : "nanjing" }
{ "_id" : ObjectId("5821d12dbf9520f3f8bfd1e7"), "tep" : 16, "location" : "nanjing" }
{ "_id" : ObjectId("5821d12dbf9520f3f8bfd1e8"), "tep" : 17, "location" : "nanjing" }
{ "_id" : ObjectId("5821d12dbf9520f3f8bfd1e9"), "tep" : 18, "location" : "nanjing" }
Type "it" for more
>
```

图 31-4　简单遍历

5. 复杂遍历

复杂遍历如图 31-5 所示。

```
> var c=db.weather.find();
>while(c.hasNext()){printjson(c.next())};
```

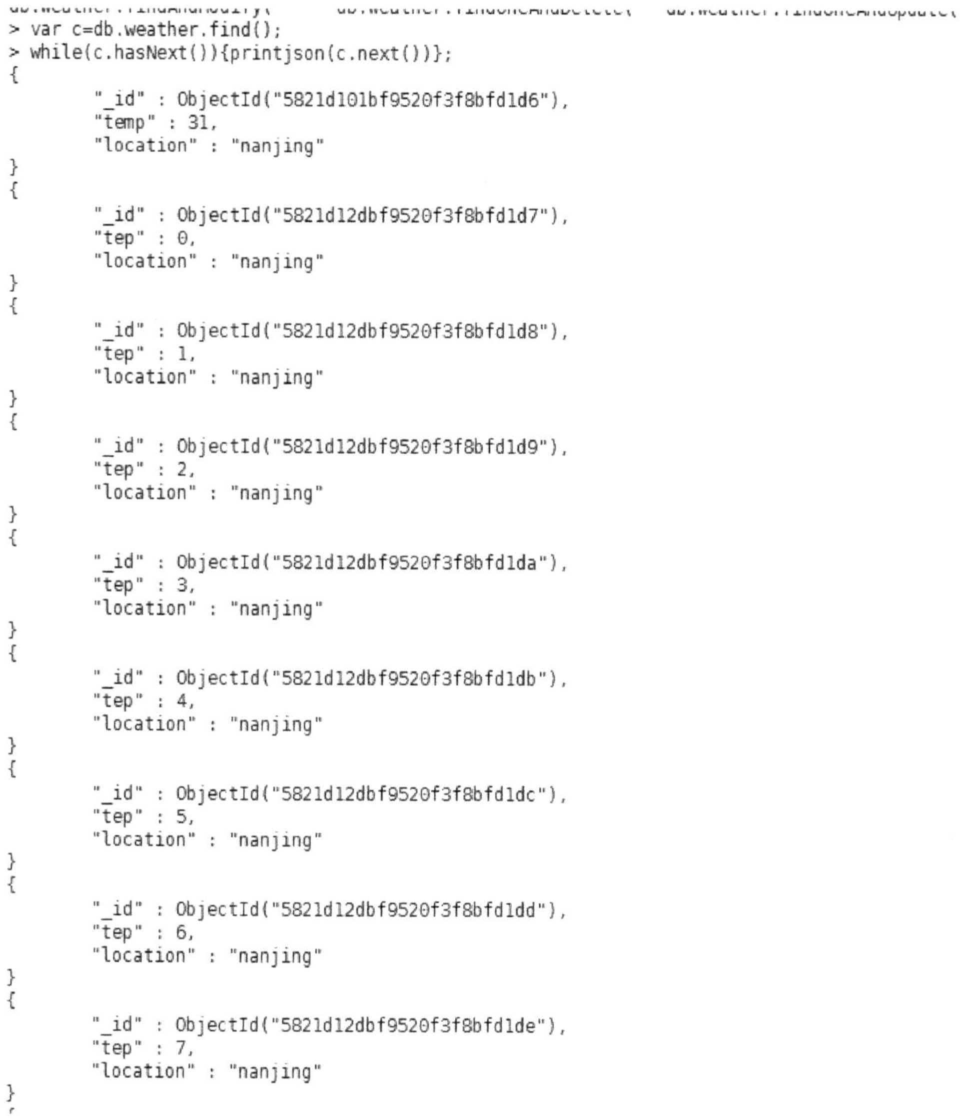

```
> var c=db.weather.find();
> while(c.hasNext()){printjson(c.next())};
{
        "_id" : ObjectId("5821d101bf9520f3f8bfd1d6"),
        "temp" : 31,
        "location" : "nanjing"
}
{
        "_id" : ObjectId("5821d12dbf9520f3f8bfd1d7"),
        "tep" : 0,
        "location" : "nanjing"
}
{
        "_id" : ObjectId("5821d12dbf9520f3f8bfd1d8"),
        "tep" : 1,
        "location" : "nanjing"
}
{
        "_id" : ObjectId("5821d12dbf9520f3f8bfd1d9"),
        "tep" : 2,
        "location" : "nanjing"
}
{
        "_id" : ObjectId("5821d12dbf9520f3f8bfd1da"),
        "tep" : 3,
        "location" : "nanjing"
}
{
        "_id" : ObjectId("5821d12dbf9520f3f8bfd1db"),
        "tep" : 4,
        "location" : "nanjing"
}
{
        "_id" : ObjectId("5821d12dbf9520f3f8bfd1dc"),
        "tep" : 5,
        "location" : "nanjing"
}
{
        "_id" : ObjectId("5821d12dbf9520f3f8bfd1dd"),
        "tep" : 6,
        "location" : "nanjing"
}
{
        "_id" : ObjectId("5821d12dbf9520f3f8bfd1de"),
        "tep" : 7,
        "location" : "nanjing"
}
```

图 31-5 复杂遍历

6. 查找数据

查找数据如图 31-6 所示。

```
> db.weather.find({location:"najing"})
```

```
> db.weather.find({location:"nanjing"});
{ "_id" : ObjectId("5821d101bf9520f3f8bfd1d6"), "temp" : 31, "location" : "nanjing" }
{ "_id" : ObjectId("5821d12dbf9520f3f8bfd1d7"), "tep" : 0, "location" : "nanjing" }
{ "_id" : ObjectId("5821d12dbf9520f3f8bfd1d8"), "tep" : 1, "location" : "nanjing" }
{ "_id" : ObjectId("5821d12dbf9520f3f8bfd1d9"), "tep" : 2, "location" : "nanjing" }
{ "_id" : ObjectId("5821d12dbf9520f3f8bfd1da"), "tep" : 3, "location" : "nanjing" }
{ "_id" : ObjectId("5821d12dbf9520f3f8bfd1db"), "tep" : 4, "location" : "nanjing" }
{ "_id" : ObjectId("5821d12dbf9520f3f8bfd1dc"), "tep" : 5, "location" : "nanjing" }
{ "_id" : ObjectId("5821d12dbf9520f3f8bfd1dd"), "tep" : 6, "location" : "nanjing" }
{ "_id" : ObjectId("5821d12dbf9520f3f8bfd1de"), "tep" : 7, "location" : "nanjing" }
{ "_id" : ObjectId("5821d12dbf9520f3f8bfd1df"), "tep" : 8, "location" : "nanjing" }
{ "_id" : ObjectId("5821d12dbf9520f3f8bfd1e0"), "tep" : 9, "location" : "nanjing" }
{ "_id" : ObjectId("5821d12dbf9520f3f8bfd1e1"), "tep" : 10, "location" : "nanjing" }
{ "_id" : ObjectId("5821d12dbf9520f3f8bfd1e2"), "tep" : 11, "location" : "nanjing" }
{ "_id" : ObjectId("5821d12dbf9520f3f8bfd1e3"), "tep" : 12, "location" : "nanjing" }
{ "_id" : ObjectId("5821d12dbf9520f3f8bfd1e4"), "tep" : 13, "location" : "nanjing" }
{ "_id" : ObjectId("5821d12dbf9520f3f8bfd1e5"), "tep" : 14, "location" : "nanjing" }
{ "_id" : ObjectId("5821d12dbf9520f3f8bfd1e6"), "tep" : 15, "location" : "nanjing" }
{ "_id" : ObjectId("5821d12dbf9520f3f8bfd1e7"), "tep" : 16, "location" : "nanjing" }
{ "_id" : ObjectId("5821d12dbf9520f3f8bfd1e8"), "tep" : 17, "location" : "nanjing" }
{ "_id" : ObjectId("5821d12dbf9520f3f8bfd1e9"), "tep" : 18, "location" : "nanjing" }
Type "it" for more
>
```

图 31-6　查找数据

7. 更新数据

更新数据如图 31-7 所示。

```
>db.weather.update({location:"nanjing"},{$set:{location:"shanghai"}},false,true);
```

```
1 master   2 master   +
> db.weather.update({location:"nanjing"},{$set:{location:"shanghai"}},false,true);
WriteResult({ "nMatched" : 31, "nUpserted" : 0, "nModified" : 31 })
>
```

图 31-7　更新数据

8. 删除数据

删除数据如图 31-8 所示。

```
> db.weather.remove({location:"nanjing"})
```

```
> db.weather.remove({location:"shanghai"});
WriteResult({ "nRemoved" : 31 })
>
```

图 31-8　删除数据

9. 删除 Collection

删除 Collection 如图 31-9 所示。

```
>db.weather.drop() ;
```

```
> db.weather.drop() ;
2016-11-08T13:29:06.542+0000 I COMMAND  [conn1] CMD: drop test.weather
true
>
```

图 31-9　删除 Collection

31.5　实验结果

本实验的实验步骤中对每一个命令都有明确截图，请读者自行进行参考。

实验三十二　LevelDB 实验：读写 LevelDB

32.1　实验目的

1. 了解 LevelDB 的使用场景；
2. 理解 LevelDB 数据存储结构；
3. 比较 LevelDB 和 Redis 的区别；
4. 对 LevelDB 的整体架构有一定了解；
5. 能正确使用 LevelDB 并能进行简单使用。

32.2　实验要求

本实验要求同学能够使用 C++语言完成对 LevelDB 库完成以下操作：
1. 连接到 LevelDB 数据库；
2. 写入数据；
3. 读取数据；
4. 删除数据。

32.3　实验原理

LevelDB 是 Google 开源的持久化 KV 单机数据库，具有很高的随机写，顺序读/写性能，但是随机读的性能很一般，也就是说，LevelDB 很适合应用在查询较少，而写很多的场景。LevelDB 应用了 LSM（Log Structured Merge）策略，lsm_tree 对索引变更进行延迟及批量处理，并通过一种类似于归并排序的方式高效地将更新迁移到磁盘，降低索引插入开销。

特点：

（1）Key 和 Value 都是任意长度的字节数组；

（2）entry（一条 K-V 记录）默认是按照 Key 的字典顺序存储的，当然开发者也可以重载这个排序函数；

（3）提供的基本操作接口：Put()、Delete()、Get()、Batch()；

（4）支持批量操作以原子操作进行；

（5）可以创建数据全景的 snapshot（快照），并允许在快照中查找数据；

（6）可以通过前向（后向）迭代器遍历数据（迭代器会隐含地创建一个 snapshot）；

（7）自动使用 Snappy 压缩数据；

（8）可移植性。

限制：

（1）非关系型数据模型（NoSQL），不支持 sql 语句，也不支持索引；

（2）一次只允许一个进程访问一个特定的数据库；

（3）没有内置的 C/S 架构，但开发者可以使用 LevelDB 库自己封装一个 server。

整体架构：LevelDB 作为存储系统，数据记录的存储介质包括内存以及磁盘文件，如果像上面说的，当 LevelDB 运行了一段时间，此时我们给 LevelDB 进行透视拍照，那么会看到如图 32-1 所示的结构。

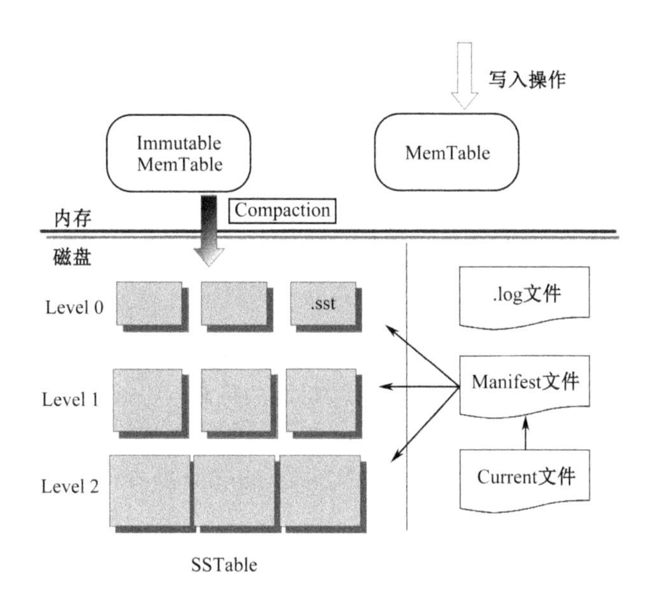

图 32-1　LevelDB 结构

从图 32-1 中可以看出，构成 LevelDB 静态结构的包括六个主要部分：内存中的 MemTable 和 Immutable MemTable 以及磁盘上的几种主要文件：Current 文件，Manifest 文件，log 文件以及 SSTable 文件。当然，LevelDB 除了这六个主要部分还有一些辅助的文件，但是以上六个文件和数据结构是 LevelDB 的主体构成元素。

LevelDB 的 Log 文件和 Memtable 与 Bigtable 论文中介绍的是一致的，当应用写入一条 Key：Value 记录的时候，LevelDB 会先往 log 文件里写入，成功后将记录插进 Memtable 中，这样基本就算完成了写入操作，因为一次写入操作只涉及一次磁盘顺序写和一次内存写入，所以这是为何说 LevelDB 写入速度极快的主要原因。

Log 文件在系统中的作用主要是用于系统崩溃恢复而不丢失数据，假如没有 Log 文件，因为写入的记录刚开始是保存在内存中的，此时如果系统崩溃，内存中的数据还没有来得及 Dump 到磁盘，所以会丢失数据（Redis 就存在这个问题）。为了避免这种情况，LevelDB 在写入内存前先将操作记录到 Log 文件中，然后再记入内存中，这样即使系统崩溃，也可以从 Log 文件中恢复内存中的 Memtable，不会造成数据的丢失。

当 Memtable 插入的数据占用内存到了一个界限后，需要将内存的记录导出到外存文件中，LevelDB 会生成新的 Log 文件和 Memtable，原先的 Memtable 就成为 Immutable Memtable，顾名思义，就是说这个 Memtable 的内容是不可更改的，只能读不能写入或者删除。新到来的数据被记入新的 Log 文件和 Memtable，LevelDB 后台调度会将 Immutable Memtable 的数据导出到磁盘，形成一个新的 SSTable 文件。SSTable 就是由内存中的数据不断导出并进行 Compaction 操作后形成的，而且 SSTable 的所有文件是一种层级结构，第一层为 Level 0，第二层为 Level 1，依次类推，层级逐渐增高，这也是称之为 LevelDB 的原因。

SSTable 中的文件是 Key 有序的，就是说在文件中小 key 记录排在大 Key 记录之前，各个 Level 的 SSTable 都是如此，但是这里需要注意的一点是：Level 0 的 SSTable 文件（后缀为.sst）和其他 Level 的文件相比有特殊性：这个层级内的.sst 文件，两个文件可能存在 key 重叠，比如有两个 level 0 的 sst 文件，文件 A 和文件 B，文件 A 的 Key 范围是：{bar, car}，文件 B 的 Key 范围是{blue,samecity}，那么很可能两个文件都存在 Key=“blood”的记录。对于其他 Level 的 SSTable 文件来说，则不会出现同一层级内.sst 文件的 Key 重叠现象，就是说 Level L 中任意两个.sst 文件，那么可以保证它们的 Key 值是不会重叠的。这点需要特别注意，后面您会看到很多操作的差异都是由于这个原因造成的。

SSTable 中的某个文件属于特定层级，而且其存储的记录是 Key 有序的，那么必然有文件中的最小 Key 和最大 Key，这是非常重要的信息，LevelDB 应该记下这些信息。Manifest 就是干这个的，它记载了 SSTable 各个文件的管理信息，比如 Level 值，文件名称，最小 Key 值和最大 Key 值。如图 32-2 是 Manifest 所存储内容的示意。

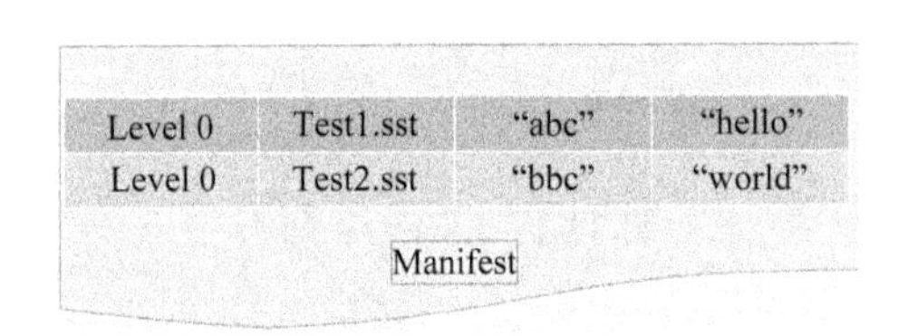

Level 0	Test1.sst	“abc”	“hello”
Level 0	Test2.sst	“bbc”	“world”

图 32-2　Manifest 存储

图 32-2 中只显示了两个文件（Manifest 会记载所有 SSTable 文件的这些信息），Level 0 的 test.sst1 和 test.sst2 文件，同时记载了这些文件各自对应的 Key 范围，例如，test.sst1 的 Key 范围是“an”到“banana”，而文件 test.sst2 的 Key 范围是“baby”到“samecity”，可以看出两者的 Key 范围是有重叠的。

Current 文件是干什么的呢？这个文件的内容只有一个信息，就是记载当前的 Manifest 文件名。因为在 LevleDB 的运行过程中，随着 Compaction 的进行，SSTable 文件会发生变化，会有新的文件产生，旧的文件被废弃，Manifest 也会跟着反映这种变化，此时往往会新生成 Manifest 文件来记载这种变化，而 Current 则用来指出哪个 Manifest 文件才是我们关心的那个 Manifest 文件。

以上介绍的内容就构成了 LevelDB 的整体静态结构。

32.4　实验步骤

在 client 机上操作：首先进入 LevelDB 目录。

```
[root@client ~]# cd /usr/cstor/leveldb
[root@client leveldb]#
```

然后创建一个 code 的文件夹作为代码编写的目录。

```
[root@client leveldb]# mkdir code
```

进入 code 目录。

```
[root@client leveldb# cd code
[root@client code]#
```

利用 vim 编写 leveldb.cpp 代码（请参考 32.4.6 小节），

编译 leveldb.cpp，编译命令如下：

```
[root@client code]# g++ -o leveldb leveldb.cpp ../out-static/libleveldb.a -lpthread -I../include
```

32.4.1　使用 C++代码建立数据库连接

核心 C++代码如下：

```
// 打开数据库连接
leveldb::Status status = leveldb::DB::Open(options,"./test_level_db", &db);
assert(status.ok());
string key = "weather";
string value = "clearday";
```

32.4.2　写入数据

写入数据的核心 C++代码如下：

```
status = db->Put(leveldb::WriteOptions(), key, value);
assert(status.ok());
```

32.4.3　读取数据

读取数据的核心 C++代码如下：

```
status = db->Get(leveldb::ReadOptions(), key, &value);
assert(status.ok());
cout<<"value :"<<value<<endl;
```

32.4.4　删除数据

删除数据的核心 C++代码如下：

```
status = db->Delete(leveldb::WriteOptions(), key);
assert(status.ok());
status = db->Get(leveldb::ReadOptions(),key, &value);
if(!status.ok()) {
    cerr<<key<<"    "<<status.ToString()<<endl;
```

```
} else {
    cout<<key<<"==="<<value<<endl;
}
```

32.4.5 关闭连接

关闭连接的核心 C++代码如下：

```
delete db;
```

32.4.6 完整的代码

```
#include <iostream>
#include <string>
#include <assert.h>
#include "leveldb/db.h"
using namespace std;
int main(void)
{
    leveldb::DB          *db;
    leveldb::Options    options;
    options.create_if_missing = true;

    // 打开数据库连接
    leveldb::Status status = leveldb::DB::Open(options,"./test_level_db", &db);
    assert(status.ok());
    string key = "weather";
    string value = "clearday";
    // 写入数据
    status = db->Put(leveldb::WriteOptions(), key, value);
    assert(status.ok());
    // 读取数据
    status = db->Get(leveldb::ReadOptions(), key, &value);
    assert(status.ok());
    cout<<"value :"<<value<<endl;
    // 删除数据
    status = db->Delete(leveldb::WriteOptions(), key);
    assert(status.ok());
    status = db->Get(leveldb::ReadOptions(),key, &value);
    if(!status.ok()) {
        cerr<<key<<"      "<<status.ToString()<<endl;
    } else {
        cout<<key<<"==="<<value<<endl;
    }
    // 关闭连接
    delete db;
```

```
    return 0;
}
```

32.5　实验结果

输入./leveldb 运行程序查看结果，如图 32-3 所示。

```
[root@client code]# ./leveldb
```

```
[root@client code]# ./leveldb
value :clearday
weather   NotFound:
[root@client code]#
```

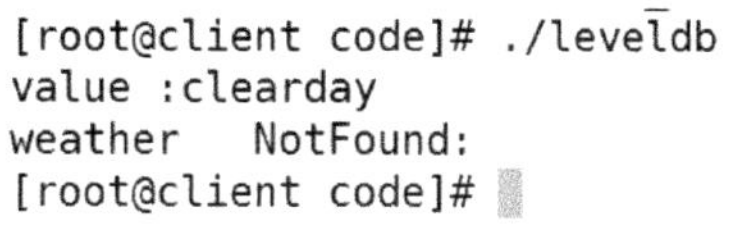

图 32-3　实验结果（一）

用 ls 命令查看 test_level_db 目录，如图 32-4 所示。

```
[root@client code]# ls test_level_db
```

```
[root@client code]# ls test_level_db
000005.ldb  000006.log  CURRENT  LOCK  LOG  LOG.old  MANIFEST-000004
[root@client code]#
```

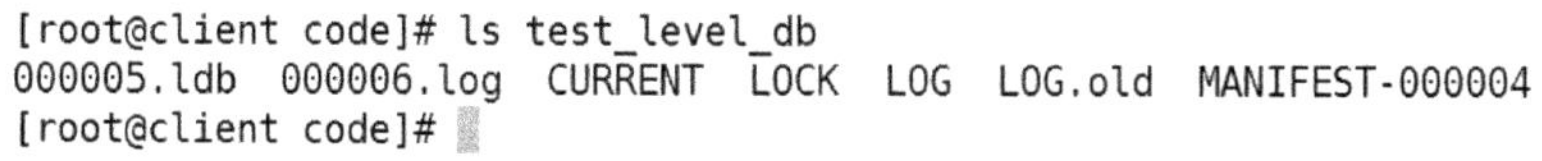

图 32-4　实验结果（二）

实验三十三　Mahout 实验：K-Means

33.1　实验目的

1．了解 Mahout 是什么；
2．了解 Mahout 能够做什么；
3．学会启动 Mahout；
4．能够通过提交 MapReduce 程序进行 K-Means 实验。

33.2　实验要求

要求实验结束时，每位学生能够在 Hadoop 集群上利用 Mahout 提交 K-Means 程序，并得出正确的实验结果。

33.3　实验原理

33.3.1　Mahout 简介

Apache Mahout 是 Apache Software Foundation（ASF）开发的一个全新的开源项目，其主要目标是创建一些可伸缩的机器学习算法，供开发人员在 Apache 在许可下免费使用。该项目已经发展到了它的第二个年头，目前只有一个公共发行版。Mahout 包含许多实现，包括集群、分类、CP 和进化程序。此外，通过使用 Apache Hadoop 库，Mahout 可以有效地扩展到云中。

33.3.2　Mahout 发展

Mahout 项目是由 Apache Lucene（开源搜索）社区中对机器学习感兴趣的一些成员发起的，他们希望建立一个可靠、文档翔实、可伸缩的项目，在其中实现一些常见的用于集群和分类的机器学习算法。该社区最初基于 Ng et al. 的文章“Map-Reduce for Machine Learning on Multicore”，但此后在发展中又并入了更多广泛的机器学习方法。Mahout 的目标还包括：

（1）建立一个用户和贡献者社区，使代码不必依赖于特定贡献者的参与或任何特定公司和大学的资金。

（2）专注于实际用例，这与高新技术研究及未经验证的技巧相反。

（3）提供高质量文章和示例。

33.3.3 Mahout 特性

虽然在开源领域中相对较为年轻，但 Mahout 已经提供了大量功能，特别是在集群和 CF 方面。Mahout 的主要特性包括：

（1）Taste CF。Taste 是 Sean Owen 在 SourceForge 上发起的一个针对 CF 的开源项目，并在 2008 年被赠予 Mahout。

（2）一些支持 Map-Reduce 的集群实现包括 K-Means、模糊 K-Means、Canopy、Dirichlet 和 Mean-Shift。

（3）Distributed Naive Bayes 和 Complementary Naive Bayes 分类实现。

（4）针对进化编程的分布式适用性功能。

（5）Matrix 和矢量库。

本实验主要介绍 K-Means。

33.3.4 K-Means 算法概要

K-Means（k-均值）是一种基于距离的聚类算法，它用质心（centroid）到属于该质的点距离这个度量来实现聚类，通常可以用于 *N* 维空间中对象。

这个算法其实很简单，如图 33-1 所示。

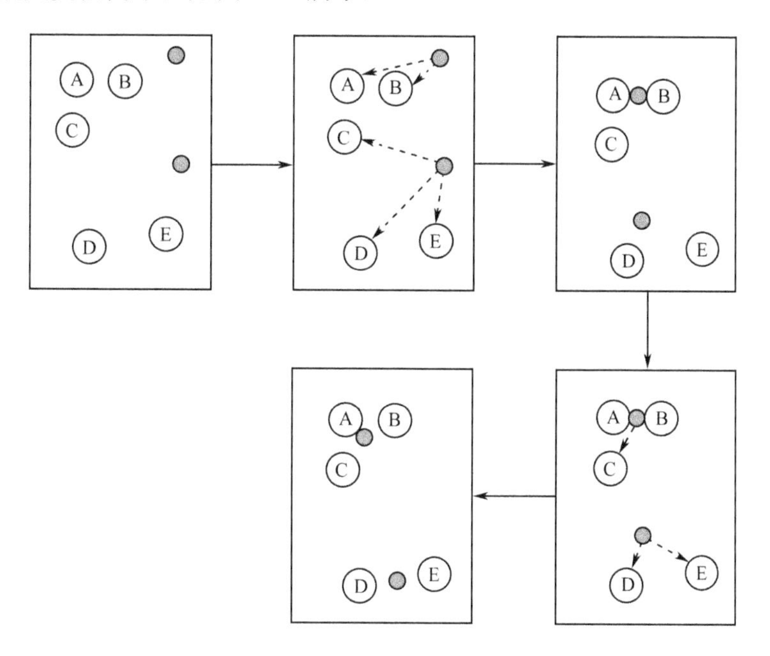

图 33-1 K-Means 算法图示

从图 33-1 中，我们可以看到，A、B、C、D、E 是五个在图中点。而灰色的点是我们的种子点，也就是我们用来找点群的点。有两个种子点，所以 K=2。

然后，K-Means 的算法如下：

（1）随机在图中取 K（这里 K=2）个种子点。

（2）然后对图中的所有点求到这 K 个种子点的距离，假如点 Pi 离种子点 Si 最近，那么 Pi 属于 Si 点群（图 33-1 中，我们可以看到 A、B 属于上面的种子点，C、D、E 属于下面中部的种子点）。

（3）接下来，我们要移动种子点到属于它的“点群”的中心（见图上的第三步）。

（4）然后重复第（2）和第（3）步，直到，种子点没有移动（我们可以看到图中的第四步上面的种子点聚合了 A、B、C，下面的种子点聚合了 D、E）。

33.3.5 K-Means 算法存在的问题

K-Means 算法的特点——采用两阶段反复循环过程算法，结束的条件是不再有数据元素被重新分配。

指定聚类：指定数据到某一个聚类，使得它与这个聚类中心的距离比它到其他聚类中心的距离要近。

修改聚类中心：

优点：本算法确定的 K 个划分到达平方误差最小。当聚类是密集的，且类与类之间区别明显时，效果较好。对于处理大数据集，这个算法是相对可伸缩和高效的，计算的复杂度为 O（NKt），其中 N 是数据对象的数目，t 是迭代的次数。一般来说，$K<<N$，$t<<N$ 。

33.3.6 K-Means 算法优点

K-Means 聚类算法的优点主要集中在:

（1）算法快速、简单；

（2）对大数据集有较高的效率并且是可伸缩性的；

（3）时间复杂度近于线性，而且适合挖掘大规模数据集。K-Means 聚类算法的时间复杂度是 O（nkt），其中 n 代表数据集中对象的数量，t 代表算法迭代的次数，k 代表簇的数目。

33.3.7 K-Means 算法缺点

在 K-Means 算法中 K 是事先给定的，这个 K 值的选定是非常难以估计的。很多时候，事先并不知道给定的数据集应该分成多少个类别才是最合适的。这也是 K-Means 算法的一个不足。有的算法是通过类的自动合并和分裂，得到较为合理的类型数目 K，如 ISODATA 算法。关于 K-Means 算法中聚类数目 K 值的确定在文献中，是根据方差分析理论，应用混合 F 统计量来确定最佳分类数，并应用了模糊划分熵来验证最佳分类数的正确性。在文献中，使用了一种结合全协方差矩阵的 RPCL 算法，并逐步删除那些只包含少量训练数据的类。而文献中使用的是一种称为次胜者受罚的竞争学习规则，来自动决定类的适当数目。它的思想是：对每个输入而言，不仅竞争获胜单元的权值被修正以适应输入值，而且对次胜单元采用惩罚的方法使之远离输入值。

在 K-Means 算法中，首先需要根据初始聚类中心来确定一个初始划分，其次对初始划分进行优化。这个初始聚类中心的选择对聚类结果有较大的影响，一旦初始值选择得不好，可能无法得到有效的聚类结果，这也成为 K-Means 算法的一个主要问题。对于该问题的解决，许多算法采用遗传算法（GA），例如，文献中采用遗传算法（GA）进行初始化，以内部聚类准则作为评价指标。

从 K-Means 算法框架可以看出，该算法需要不断地进行样本分类调整，不断地计算调整后的新的聚类中心，因此当数据量非常大时，算法的时间开销是非常大的。所以需要对算法的时间复杂度进行分析、改进，提高算法应用范围。在文献中从该算法的时间复杂度进行分析考虑，通过一定的相似性准则来去掉聚类中心的侯选集。而在文献中，使用的 K-Means 算法是对样本数据进行聚类，无论是初始点的选择还是一次迭代完成时对数据的调整，都是建立在随机选取的样本数据的基础之上，这样可以提高算法的收敛速度。

33.3.8 K-Means 算法应用

看到这里，你会说 K-Means 算法看来很简单，而且好像就是在玩坐标点，没什么真实用处。而且，这个算法缺陷很多，还不如人工呢。是的，前面的例子只是玩二维坐标点，的确没什么意思。但是你想一想下面的几个问题：

（1）如果不是二维的，是多维的，如五维的，那么，就只能用计算机来计算了。

（2）二维坐标点的 *X*、*Y* 坐标，其实是一种向量，是一种数学抽象。现实世界中很多属性是可以抽象成向量的，例如，我们的年龄、喜好、商品等，能抽象成向量的目的就是可以让计算机知道某两个属性间的距离。

只要能把现实世界的物体的属性抽象成向量，就可以用 K-Means 算法来归类了。

33.4 实验步骤

33.4.1 添加临时 JAVA_HOME 环境变量

添加临时 JAVA_HOME 环境变量。

```
[root@client hadoop]# export JAVA_HOME=/usr/local/jdk1.7.0_79
```

33.4.2 建立 HDFS 目录

在 client 机上操作：首先在 HDFS 上建立目录。

```
[root@client hadoop]# cd /usr/cstor/hadoop/
[root@client hadoop]# bin/hadoop fs -mkdir -p /usr/cstor/hadoop/testdata
```

33.4.3 实验数据准备

其次，将/root/data/33/文件夹下的 *synthetic_control.data* 文件上传到 HDFS 上，放在刚刚新建好的目录下面。

```
[root@client hadoop]# bin/hadoop fs -put /root/data/33/synthetic_control.data /usr/cstor/hadoop/testdata
```

33.4.4 提交 Mahout 的 K-Means 程序

执行代码提交命令，提交 Mahout 的 K-Means 程序。

```
[root@client hadoop]# bin/hadoop jar /usr/cstor/mahout/mahout-examples-0.9-job.jar \
> org.apache.mahout.clustering.syntheticcontrol.kmeans.Job
```

33.5 实验结果

在 client 上执行对 HDFS 上的文件夹/usr/cstor/hadoop/output 内容查看的操作。

```
[root@client hadoop]# bin/hadoop fs -ls /usr/cstor/hadoop/output/
```

屏幕上显示如图 33-2 所示的内容。

```
[root@client hadoop]# bin/hadoop fs -ls /usr/cstor/hadoop/output/
17/02/07 12:00:39 WARN util.NativeCodeLoader: Unable to load native-hadoop library for your platform.
Found 15 items
-rw-r--r--   1 root root        194 2017-02-07 11:43 /usr/cstor/hadoop/output/_policy
drwxr-xr-x   - root root         84 2017-02-07 11:43 /usr/cstor/hadoop/output/clusteredPoints
drwxr-xr-x   - root root       4096 2017-02-07 11:43 /usr/cstor/hadoop/output/clusters-0
drwxr-xr-x   - root root        117 2017-02-07 11:43 /usr/cstor/hadoop/output/clusters-1
drwxr-xr-x   - root root        117 2017-02-07 11:43 /usr/cstor/hadoop/output/clusters-10-final
drwxr-xr-x   - root root        117 2017-02-07 11:43 /usr/cstor/hadoop/output/clusters-2
drwxr-xr-x   - root root        117 2017-02-07 11:43 /usr/cstor/hadoop/output/clusters-3
drwxr-xr-x   - root root        117 2017-02-07 11:43 /usr/cstor/hadoop/output/clusters-4
drwxr-xr-x   - root root        117 2017-02-07 11:43 /usr/cstor/hadoop/output/clusters-5
drwxr-xr-x   - root root        117 2017-02-07 11:43 /usr/cstor/hadoop/output/clusters-6
drwxr-xr-x   - root root        117 2017-02-07 11:43 /usr/cstor/hadoop/output/clusters-7
drwxr-xr-x   - root root        117 2017-02-07 11:43 /usr/cstor/hadoop/output/clusters-8
drwxr-xr-x   - root root        117 2017-02-07 11:43 /usr/cstor/hadoop/output/clusters-9
drwxr-xr-x   - root root         84 2017-02-07 11:43 /usr/cstor/hadoop/output/data
drwxr-xr-x   - root root         55 2017-02-07 11:43 /usr/cstor/hadoop/output/random-seeds
[root@client hadoop]#
```

图 33-2　查看文件列表

实验三十四　使用 Spark 实现 K-Means

34.1　实验目的

1. 熟练使用 Spark Shell 接口；
2. 了解 K-Means 算法原理；
3. 理解 K-Means 执行过程；
4. 配置 Spark 处理 HDFS 数据；
5. 使用 Spark 机器学习包中 K-Means 工具包处理 HDFS 中数据。

34.2　实验要求

使用 Spark MLlib（机器学习库）中的 K-Means 工具包，对存储在 HDFS 上的数据集 sample_kmeans_data.txt 进行聚类。

34.3　实验原理

请参考实验三十三实验原理。

34.4　实验步骤

K-Means 主要用于在对样本类别不了解情况下，对样本进行聚类，其本身无“训练、预测”这一说法，直接设置中心点数量或终止距离即可。下面将按照“准备数据集→使用 K-Means 进行聚类”依次讲述。

34.4.1　添加临时 JAVA_HOME 环境变量

首先，手工或者使用“一键搭建 spark”功能构筑好 Spark 环境。

登录 slave1 机，添加临时 JAVA_HOME 环境变量。

```
[root@slave1 ~]# export JAVA_HOME=/usr/local/jdk1.7.0_79
```

34.4.2　上传训练数据集

查看 HDFS 里是否已存在目录“/34/in”，若不存在，使用下述命令新建该目录（见图 34-1）。

```
[root@slave1 ~]# /usr/cstor/hadoop/bin/hdfs dfs -mkdir -p /34/in
```

```
[root@slave1 ~]# /usr/cstor/hadoop/bin/hdfs dfs -mkdir -p /34/in
17/02/13 09:36:40 WARN util.NativeCodeLoader: Unable to load native-hadoop library for your platform... using builtin-java classes where applicable
```

图 34-1 创建目录

其次，使用下述命令将 slave1 机本地文件“/usr/cstor/spark/data/mllib/ kmeans_data.txt”上传至 HDFS 的“/34/in”目录（见图 34-2）。

```
[root@slave1 ~]# /usr/cstor/hadoop/bin/hdfs dfs -put /usr/cstor/spark/data/mllib/kmeans_data.txt /34/in
```

```
[root@slave1 ~]# /usr/cstor/hadoop/bin/hdfs dfs -put /usr/cstor/spark/data/mllib/kmeans_data.txt /34/in
17/02/13 09:36:53 WARN util.NativeCodeLoader: Unable to load native-hadoop library for your platform... using builtin-java classes where applicable
```

图 34-2 上传文件

最后，请确认 HDFS 上存在文件“/usr/cstor/spark/data/mllib/kmeans_data.txt”（见图 34-3）。

```
[root@slave1 ~]# /usr/cstor/hadoop/bin/hdfs dfs -ls /34/in
17/02/13 10:04:44 WARN util.NativeCodeLoader: Unable to load native-hadoop library for your platform... using builtin-java classes where applicable
Found 1 items
-rw-r--r--   3 root supergroup         72 2017-02-13 09:36 /34/in/kmeans_data.txt
```

图 34-3 确认文件是否上传成功

34.4.3 训练 SVM 模型

准备好输入文件后，下一步便是在 Spark 集群上执行 K-Means 程序（处理该数据集）。下面的处理代码参考自“/usr/cstor/spark/examples/src/main/scala/org/apache/spark/examples/SparkKMeans.scala”，操作命令主要在 client 机上完成。

首先，在 slave1 机上，使用下述命令，进入 Spark Shell 接口。

```
[root@slave1 ~]# /usr/cstor/spark/bin/spark-shell --master spark://master:7077
```

进入 Spark Shell 命令行执行环境后，依次输入下述代码，完成模型训练。

```
import breeze.linalg.{Vector, DenseVector, squaredDistance}
import org.apache.spark.{SparkConf, SparkContext}
import org.apache.spark.SparkContext._
def parseVector(line: String): Vector[Double] = {
DenseVector(line.split(' ').map(_.toDouble))
}
def closestPoint(p: Vector[Double], centers: Array[Vector[Double]]): Int = {
var bestIndex = 0
var closest = Double.PositiveInfinity
for (i <- 0 until centers.length) {
val tempDist = squaredDistance(p, centers(i))
if (tempDist < closest) {
closest = tempDist
```

```
bestIndex = i
}
}
bestIndex
}
val lines = sc.textFile("/34/in/kmeans_data.txt")
val data = lines.map(parseVector _).cache()
val K = "2".toInt
val convergeDist = "0.1".toDouble
val kPoints = data.takeSample(withReplacement = false, K, 42).toArray
var tempDist = 1.0
while(tempDist > convergeDist) {
val closest = data.map (p => (closestPoint(p, kPoints), (p, 1)))
val pointStats = closest.reduceByKey{case ((p1, c1), (p2, c2)) => (p1 + p2, c1 + c2)}
val newPoints = pointStats.map {pair =>
(pair._1, pair._2._1 * (1.0 / pair._2._2))}.collectAsMap()
tempDist = 0.0
for (i <- 0 until K) {
tempDist += squaredDistance(kPoints(i), newPoints(i))
}
for (newP <- newPoints) {
kPoints(newP._1) = newP._2
}
println("Finished iteration (delta = " + tempDist + ")")
}
println("Final centers:")
kPoints.foreach(println)
```

最后，可以直接复制上述代码粘贴至在 Spark 交换式执行器中执行。

34.5　实验结果

实验直接使用 Spark 自带的 K-Means 数据集与 K-Means 代码库，省去大量代码篇幅，读者可下载源码包，然后逐个跟进便可看到完整 K-Means 处理源码。

本实验完成使用 K-Means 算法对 sample_kmeans_data.txt 内所有样本进行聚类，代码最后一行便是在控制台上打印中心点，最终执行结果如图 34-4 所示。

```
scala> println("Final centers:")
Final centers:

scala> kPoints.foreach(println)
DenseVector(0.1, 0.1, 0.1)
DenseVector(9.099999999999998, 9.099999999999998, 9.099999999999998)
```

图 34-4　实验结果

实验三十五　使用 Spark 实现 SVM

35.1　实验目的

1．熟练使用 Spark Shell 接口；
2．了解 SVM（支持向量机）算法原理；
3．理解 SVM 执行过程；
4．配置 Spark 处理 HDFS 数据；
5．使用 Spark 机器学习包中 SVM 工具包处理 HDFS 上数据。

35.2　实验要求

使用 Spark MLlib（机器学习库）中的 SVM 工具包，训练存储在 HDFS 上的 SVM 训练数据集 sample_libsvm_data.txt。

35.3　实验原理

35.3.1　SVM 算法介绍

支持向量机（Support Vector Machine，SVM）是一种分类算法，通过寻求结构风险最小化来提高学习机泛化能力，实现经验风险和置信范围的最小化，从而达到在统计样本量较少的情况下，亦能获得良好统计规律的目的。通俗来讲，它是一种两类分类模型，其基本模型定义为特征空间上的间隔最大的线性分类器，支持向量机的学习策略便是间隔最大化，最终可转化为一个凸二次规划问题的求解。

35.3.2　SVM 算法原理

（1）在 n 维空间中找到一个分类超平面，将空间上的点分类（见图 35-1）。

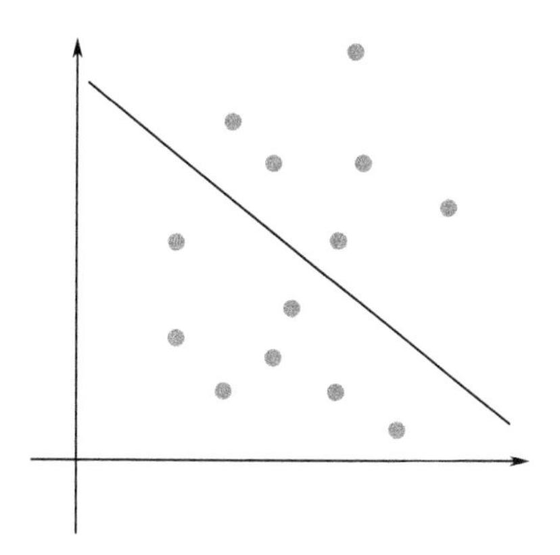

图 35-1　SVM 算法图示

（2）一般而言，一个点距离超平面的远近可以表示为分类预测的确信或准确程度。SVM 就是要最大化这个间隔值。而在虚线上的点便叫做支持向量（Supprot Verctor），如图 35-2 所示。

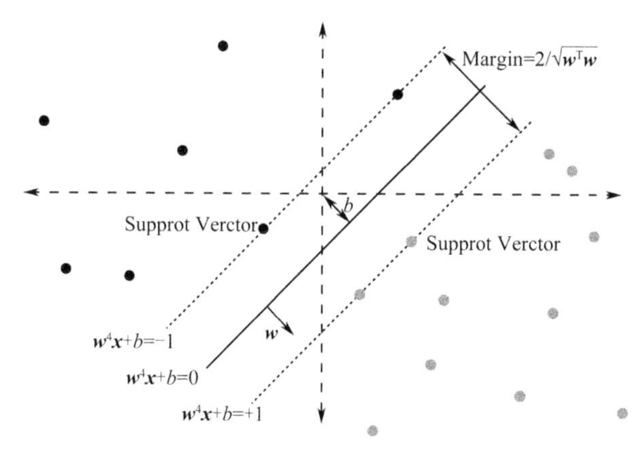

图 35-2　支持向量

（3）实际中，我们会经常遇到线性不可分的样例，此时，我们的常用做法是把样例特征映射到高维空间中去，如图 35-3 所示。

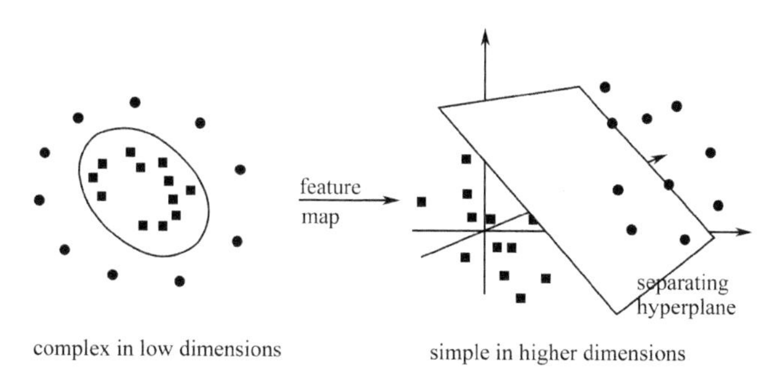

图 35-3　高维映射

（4）线性不可分映射到高维空间，可能会导致维度大小高到可怕的（19 维乃至无穷维的例子），导致计算复杂。核函数的价值在于它虽然也是讲特征进行从低维到高维的转换，但核函数绝就绝在它事先在低维上进行计算，而将实质上的分类效果表现在了高维上，也就如上文所说的避免了直接在高维空间中的复杂计算。

（5）使用松弛变量处理数据噪声。

35.4 实验步骤

当需要使用 SVM 算法对样本进行分类时，首先应基于已有样本，训练 SVM 模型，然后才可使用该模型对新样本进行预测。下面将按照“准备数据集→训练 SVM 分类器→使用 SVM 分类器”依次讲述。

35.4.1 添加临时 JAVA_HOME 环境变量

首先，手工或者使用“一键搭建 spark”功能构筑好 Spark 环境。

登录 slave1 机，添加临时 JAVA_HOME 环境变量。

```
[root@slave1 ~]# export JAVA_HOME=/usr/local/jdk1.7.0_79
```

35.4.2 上传训练数据集

查看 HDFS 里是否已存在目录“/35/in”。若不存在，使用下述命令新建该目录。

```
[root@slave1 ~]# /usr/cstor/hadoop/bin/hdfs  dfs  -mkdir  -p  /35/in
```

其次，使用下述命令将 slave1 机本地文件“/root/data/35/sample_libsvm_data.txt”上传至 HDFS 的“/35/in”目录：

```
[root@slave1 ~]# /usr/cstor/hadoop/bin/hdfs dfs  -put  /root/data/35/sample_libsvm_data.txt  /35/in
```

最后，请确认 HDFS 上已经存在文件 sample_libsvm_data.txt，如图 35-4 所示。

```
[root@slave1 ~]# /usr/cstor/hadoop/bin/hdfs dfs  -ls  /35/in
17/02/13 11:49:04 WARN util.NativeCodeLoader: Unable to load native-hadoop library for your pl
Found 2 items
-rw-r--r--   3 root supergroup     104736 2017-02-13 11:20 /35/in/sample_libsvm_data.txt
```

图 35-4 确认存在文件 sample_libsvm_data.txt

35.4.3 训练 SVM 模型

准备好输入文件后，下一步便是在 Spark 集群上执行 SVM 程序（处理该数据集）。下面的处理代码参考自“http://spark.apache.org/docs/latest/mllib-linear-methods.html”，操作命令主要在 master 机上完成。

首先，在 slave1 机上，使用下述命令，进入 Spark Shell 接口，如图 35-5 所示。

```
[root@slave1 ~]#  /usr/cstor/spark/bin/spark-shell  --master  spark://master:7077
```

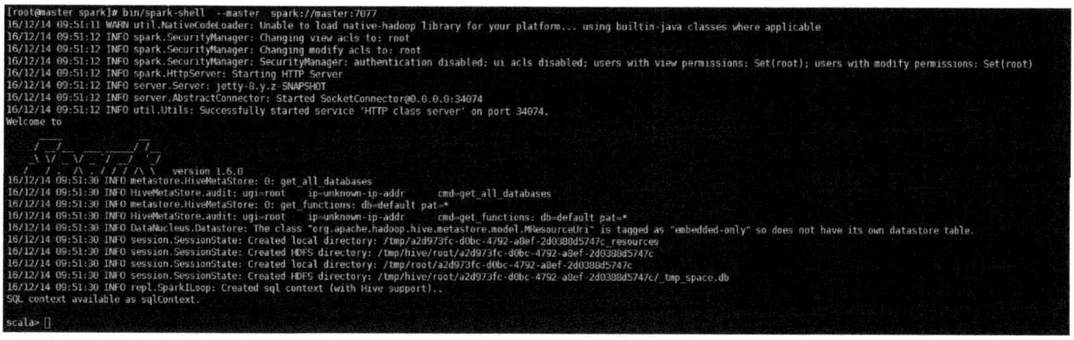

```
[root@master spark]# bin/spark-shell  --master  spark://master:7077
16/12/14 09:51:11 WARN util.NativeCodeLoader: Unable to load native-hadoop library for your platform... using builtin-java classes where applicable
16/12/14 09:51:12 INFO spark.SecurityManager: Changing view acls to: root
16/12/14 09:51:12 INFO spark.SecurityManager: Changing modify acls to: root
16/12/14 09:51:12 INFO spark.SecurityManager: SecurityManager: authentication disabled; ui acls disabled; users with view permissions: Set(root); users with modify permissions: Set(root)
16/12/14 09:51:12 INFO spark.HttpServer: Starting HTTP Server
16/12/14 09:51:12 INFO server.Server: jetty-8.y.z-SNAPSHOT
16/12/14 09:51:12 INFO server.AbstractConnector: Started SocketConnector@0.0.0.0:34074
16/12/14 09:51:12 INFO util.Utils: Successfully started service 'HTTP class server' on port 34074.
Welcome to
                                   version 1.6.0
16/12/14 09:51:30 INFO metastore.HiveMetaStore: 0: get_all_databases
16/12/14 09:51:30 INFO HiveMetaStore.audit: ugi=root    ip=unknown-ip-addr      cmd=get_all_databases
16/12/14 09:51:30 INFO metastore.HiveMetaStore: 0: get_functions: db=default pat=*
16/12/14 09:51:30 INFO HiveMetaStore.audit: ugi=root    ip=unknown-ip-addr      cmd=get_functions: db=default pat=*
16/12/14 09:51:30 INFO DataNucleus.Datastore: The class "org.apache.hadoop.hive.metastore.model.MResourceUri" is tagged as "embedded-only" so does not have its own datastore table.
16/12/14 09:51:30 INFO session.SessionState: Created local directory: /tmp/a2d973fc-d0bc-4792-a8ef-2d0388d5747c_resources
16/12/14 09:51:30 INFO session.SessionState: Created HDFS directory: /tmp/hive/root/a2d973fc-d0bc-4792-a8ef-2d0388d5747c
16/12/14 09:51:30 INFO session.SessionState: Created local directory: /tmp/root/a2d973fc-d0bc-4792-a8ef-2d0388d5747c
16/12/14 09:51:30 INFO session.SessionState: Created HDFS directory: /tmp/hive/root/a2d973fc-d0bc-4792-a8ef-2d0388d5747c/_tmp_space.db
16/12/14 09:51:30 INFO repl.SparkILoop: Created sql context (with Hive support)..
SQL context available as sqlContext.

scala>
```

图 35-5　启动 master 节点

进入 Spark Shell 命令行执行环境后，依次输入下述代码，完成模型训练。

```
import org.apache.spark.mllib.classification.{SVMModel, SVMWithSGD}
import org.apache.spark.mllib.evaluation.BinaryClassificationMetrics
import org.apache.spark.mllib.util.MLUtils
// Load training data in LIBSVM format.
val data = MLUtils.loadLibSVMFile(sc, "/35/in/sample_libsvm_data.txt")
// Split data into training (60%) and test (40%).
val splits = data.randomSplit(Array(0.6, 0.4), seed = 11L)
val training = splits(0).cache()
val test = splits(1)
// Run training algorithm to build the model
val numIterations = 100
val model = SVMWithSGD.train(training, numIterations)
// Clear the default threshold.
model.clearThreshold()
// Compute raw scores on the test set.
val scoreAndLabels = test.map { point =>
  val score = model.predict(point.features)
  (score, point.label)
}
// Get evaluation metrics.
val metrics = new BinaryClassificationMetrics(scoreAndLabels)
val auROC = metrics.areaUnderROC()
println("Area under ROC = " + auROC)
// Save and load model
model.save(sc, "/35/in/scalaSVMWithSGDModel")
val sameModel = SVMModel.load(sc, "/35/in/scalaSVMWithSGDModel")
```

35.5　实验结果

实验直接使用 Spark 自带的 SVM 数据集与 SVM 代码库，省去大量代码篇幅，读者可下载源码包，然后逐个跟进便可看到完整 SVM 处理源码。

实验结果是调用训练好的 SVM 模型对新样本进行分类，即 35.3.2 节，其执行过程如图 35-6 所示。

```
16/12/14 10:28:31 INFO spark.SparkContext: Created broadcast 221 from take at GLMClassificationModel.scala:79
16/12/14 10:28:31 INFO parquet.ParquetRelation: Reading Parquet file(s) from hdfs://master:8020/user/spark/data/mllib/scalaSVMWithSGDModel/data/part-r-00000-08ffcf4f-cd61-4e96-b249-980f31b50381.gz.parquet
16/12/14 10:28:31 INFO spark.SparkContext: Starting job: take at GLMClassificationModel.scala:79
16/12/14 10:28:31 INFO scheduler.DAGScheduler: Got job 115 (take at GLMClassificationModel.scala:79) with 1 output partitions
16/12/14 10:28:31 INFO scheduler.DAGScheduler: Final stage: ResultStage 122 (take at GLMClassificationModel.scala:79)
16/12/14 10:28:31 INFO scheduler.DAGScheduler: Parents of final stage: List()
16/12/14 10:28:31 INFO scheduler.DAGScheduler: Missing parents: List()
16/12/14 10:28:31 INFO scheduler.DAGScheduler: Submitting ResultStage 122 (MapPartitionsRDD[245] at take at GLMClassificationModel.scala:79), which has no missing parents
16/12/14 10:28:31 INFO storage.MemoryStore: Block broadcast_222 stored as values in memory (estimated size 5.7 KB, free 1020.6 KB)
16/12/14 10:28:31 INFO storage.MemoryStore: Block broadcast_222_piece0 stored as bytes in memory (estimated size 3.3 KB, free 1023.8 KB)
16/12/14 10:28:31 INFO storage.BlockManagerInfo: Added broadcast_222_piece0 in memory on 10.1.1.21:34728 (size: 3.3 KB, free: 511.3 MB)
16/12/14 10:28:31 INFO spark.SparkContext: Created broadcast 222 from broadcast at DAGScheduler.scala:1006
16/12/14 10:28:31 INFO scheduler.DAGScheduler: Submitting 1 missing tasks from ResultStage 122 (MapPartitionsRDD[245] at take at GLMClassificationModel.scala:79)
16/12/14 10:28:31 INFO scheduler.TaskSchedulerImpl: Adding task set 122.0 with 1 tasks
16/12/14 10:28:31 INFO scheduler.TaskSetManager: Starting task 0.0 in stage 122.0 (TID 371, slave3, partition 0,NODE_LOCAL, 2261 bytes)
16/12/14 10:28:31 INFO storage.BlockManagerInfo: Added broadcast_222_piece0 in memory on slave3:38539 (size: 3.3 KB, free: 511.3 MB)
16/12/14 10:28:31 INFO storage.BlockManagerInfo: Added broadcast_221_piece0 in memory on slave3:38539 (size: 19.4 KB, free: 511.3 MB)
16/12/14 10:28:32 INFO scheduler.TaskSetManager: Finished task 0.0 in stage 122.0 (TID 371) in 494 ms on slave3 (1/1)
16/12/14 10:28:32 INFO scheduler.DAGScheduler: ResultStage 122 (take at GLMClassificationModel.scala:79) finished in 0.494 s
16/12/14 10:28:32 INFO scheduler.TaskSchedulerImpl: Removed TaskSet 122.0, whose tasks have all completed, from pool
16/12/14 10:28:32 INFO scheduler.DAGScheduler: Job 115 finished: take at GLMClassificationModel.scala:79, took 0.500972 s
sameModel: org.apache.spark.mllib.classification.SVMModel = org.apache.spark.mllib.classification.SVMModel: intercept = 0.0, numFeatures = 692, numClasses = 2, threshold = None
```

图 35-6　实验结果

实验三十六　使用 Spark 实现 FP-Growth

36.1　实验目的

1. 熟练使用 Spark Shell 接口；
2. 了解 FP-Growth 关联分析原理；
3. 了解 FP-Growth 算法执行过程；
4. 配置 Spark 处理 HDFS 数据；
5. 使用 Spark 机器学习包中 FP-Growth 算法处理 HDFS 中数据。

36.2　实验要求

使用 Spark MLlib（机器学习库）中的 FP-Growth 算法，处理存储在 HDFS 上的数据集 sample_fpgrowth.txt。

36.3　实验原理

36.3.1　FP-Growth 算法简介

FP 的全称是 Frequent Pattern，在算法中使用了一种称为频繁模式树（Frequent Pattern Tree）的数据结构。FP trec 是一种特殊的前缀树，由频繁项头表和项前缀树构成。所谓前缀树，是一种存储候选项集的数据结构，树的分支用项名标识，树的节点存储后缀项，路径表示项集。

FP-tree 的生成方法如图 36-1 所示。

根据支持度对频繁项进行排序是本算法的关键。一是通过将支持度高的项排在前面，使得生成的 FP-tree 中，出现频繁的项更可能被共享，从而有效地节省算法运行所需要的空间。二是通过这种排序，可以对 FP-tree 所包含的频繁模式进行互斥的空间拆分，得到相互独立的子集，而这些子集又组成了完整的信息。

FP-tree 子集分割方法：

求 p 为前缀的投影数据库：根据头表的指针找到 FP-tree 的两个 p 节点，搜索出从这两个节点到树的根节点路径节点信息（包含支持度）。然后累加路径节点信息的支持度，删除非频繁项。对剩下的频繁项按照上一节的方法构建 FP-tree。如图 36-2 和图 36-3 所示。

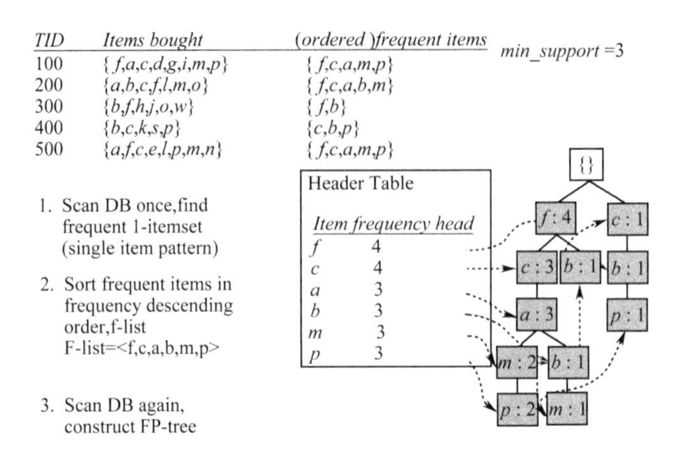

图 36-1 FP-tree 的生成方法

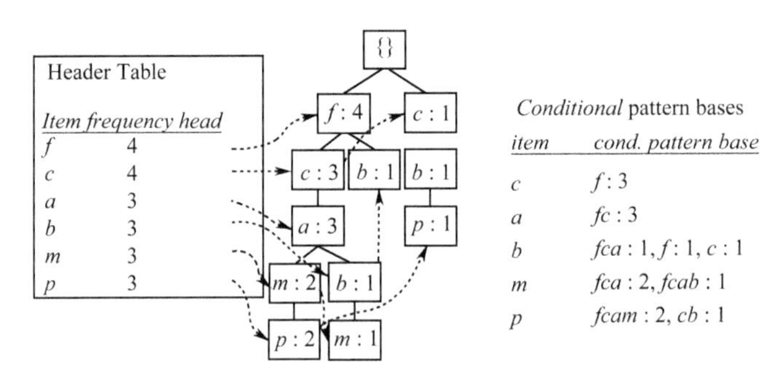

图 36-2 搜索节点信息

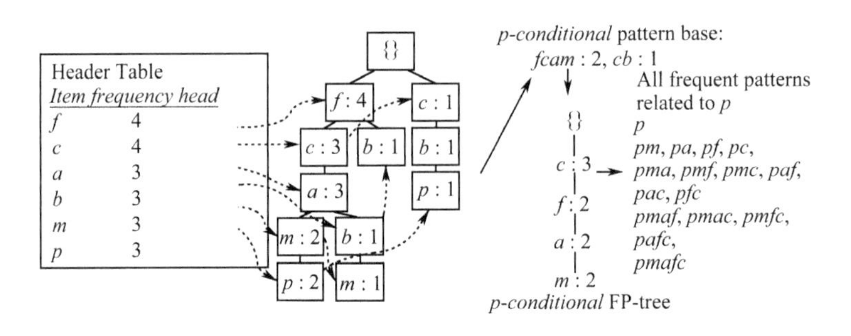

图 36-3 构建 FP-tree

36.3.2 FP-Growth 算法流程

不断地迭代 FP-tree 的构造和投影过程。

对于每个频繁项，构造它的条件投影数据库和投影 FP-tree。对每个新构建的 FP-tree 重复这个过程，直到构造新的 FP-tree 为空，或者只包含一条路径。当构造的 FP-tree 为空时，其前缀为频繁模式；当只包含一条路径时，通过枚举所有可能组合并与此树的前缀连接即可得到频繁模式。

36.4 实验步骤

关联分析指的是从大规模数据集中寻找物品间的隐含关系，这些关系主要包括“频繁项集”和“关联规则”，挖掘出“频繁项集”后，可从中找出“关联规则”。当使用 FP-Growth 算法寻找频繁项时，首先需要构建 FP 树，其次可通过查找元素项的条件基、构建条件 FP 树来发现频繁项集，最后再从该“频繁项集”中计算出“关联规则”。下述步骤即讲述这一过程。

36.4.1 添加临时 JAVA_HOME 环境变量

首先，手工或者使用“一键搭建 spark”功能构筑好 Spark 环境。

登录 slave1 机，添加临时 JAVA_HOME 环境变量。

```
[root@slave1 ~]# export JAVA_HOME=/usr/local/jdk1.7.0_79
```

36.4.2 上传训练数据集

查看 HDFS 里是否已存在目录“/36/in”。若不存在，使用下述命令新建该目录。

```
[root@slave1 ~]# /usr/cstor/hadoop/bin/hdfs  dfs  -mkdir  -p  /36/in
```

其次，使用下述命令将 slave1 机本地文件“/root/data/36/sample_fpgrowth.txt”上传至 HDFS 的“/36/in”目录。

```
[root@slave1 ~]# /usr/cstor/hadoop/bin/hdfs dfs  -put  /root/data/36/sample_fpgrowth.txt  /36/in
```

最后，请确认 HDFS 上已经存在文件 sample_fpgrowth.txt。

```
[root@slave1 ~]# /usr/cstor/hadoop/bin/hdfs  dfs  -ls  /36/in
17/02/13 11:58:19 WARN util.NativeCodeLoader: Unable to load native-hadoop library for
Found 1 items
-rw-r--r--   3 root supergroup         68 2017-02-13 11:58 /36/in/sample_fpgrowth.txt
```

36.4.3 训练 SVM 模型

准备好输入文件后，下一步便是在 Spark 集群上执行 SVM 程序（处理该数据集）。下面的处理代码参考自“http://spark.apache.org/docs/latest/mllib-linear-methods.html”，操作命令主要在 slave1 机上完成。

首先，在 slave1 机上，使用下述命令，进入 Spark Shell 接口，如图 36-4 所示。

```
[root@slave1 ~]#  /usr/cstor/spark/bin/spark-shell  --master  spark://master:7077
```

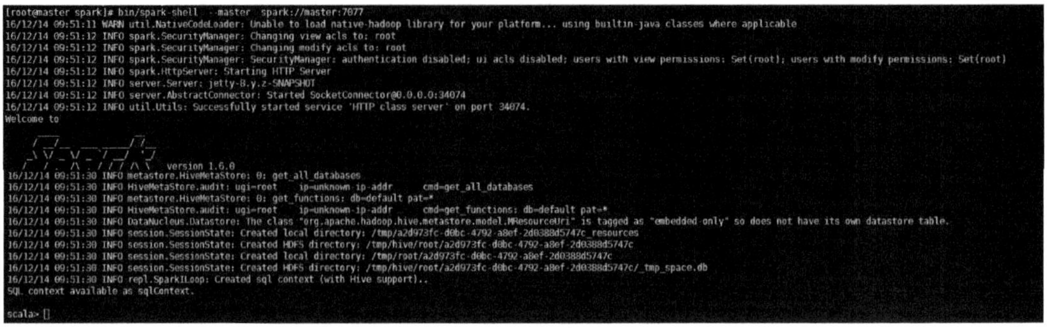

```
[root@master spark]# bin/spark-shell --master spark://master:7077
16/12/14 09:51:11 WARN util.NativeCodeLoader: Unable to load native-hadoop library for your platform... using builtin-java classes where applicable
16/12/14 09:51:12 INFO spark.SecurityManager: Changing view acls to: root
16/12/14 09:51:12 INFO spark.SecurityManager: Changing modify acls to: root
16/12/14 09:51:12 INFO spark.SecurityManager: SecurityManager: authentication disabled; ui acls disabled; users with view permissions: Set(root); users with modify permissions: Set(root)
16/12/14 09:51:12 INFO spark.HttpServer: Starting HTTP Server
16/12/14 09:51:12 INFO server.Server: jetty-8.y.z-SNAPSHOT
16/12/14 09:51:12 INFO server.AbstractConnector: Started SocketConnector@0.0.0.0:34074
16/12/14 09:51:12 INFO util.Utils: Successfully started service 'HTTP class server' on port 34074.
Welcome to
                                   version 1.6.0
16/12/14 09:51:30 INFO metastore.HiveMetaStore: 0: get_all_databases
16/12/14 09:51:30 INFO HiveMetaStore.audit: ugi=root    ip=unknown-ip-addr      cmd=get_all_databases
16/12/14 09:51:30 INFO metastore.HiveMetaStore: 0: get_functions: db=default pat=*
16/12/14 09:51:30 INFO HiveMetaStore.audit: ugi=root    ip=unknown-ip-addr      cmd=get_functions: db=default pat=*
16/12/14 09:51:30 INFO DataNucleus.Datastore: The class "org.apache.hadoop.hive.metastore.model.MResourceUri" is tagged as "embedded-only" so does not have its own datastore table.
16/12/14 09:51:30 INFO session.SessionState: Created local directory: /tmp/a2d973fc-d0bc-4792-a8ef-2d0388d5747c_resources
16/12/14 09:51:30 INFO session.SessionState: Created HDFS directory: /tmp/hive/root/a2d973fc-d0bc-4792-a8ef-2d0388d5747c
16/12/14 09:51:30 INFO session.SessionState: Created local directory: /tmp/root/a2d973fc-d0bc-4792-a8ef-2d0388d5747c
16/12/14 09:51:30 INFO session.SessionState: Created HDFS directory: /tmp/hive/root/a2d973fc-d0bc-4792-a8ef-2d0388d5747c/_tmp_space.db
16/12/14 09:51:30 INFO repl.SparkILoop: Created sql context (with Hive support)..
SQL context available as sqlContext.

scala> []
```

图 36-4 启动 master 节点

进入 Spark Shell 命令行执行环境后，依次输入下述代码，完成模型训练。

```
import org.apache.spark.mllib.fpm.FPGrowth
import org.apache.spark.rdd.RDD
val data = sc.textFile("data/mllib/sample_fpgrowth.txt")
val transactions: RDD[Array[String]] = data.map(s => s.trim.split(' '))
val fpg = new FPGrowth()
  .setMinSupport(0.2)
  .setNumPartitions(10)
val model = fpg.run(transactions)
model.freqItemsets.collect().foreach { itemset =>
  println(itemset.items.mkString("[", ",", "]") + ", " + itemset.freq)
}
val minConfidence = 0.8
model.generateAssociationRules(minConfidence).collect().foreach { rule =>
  println(
    rule.antecedent.mkString("[", ",", "]")
      + " => " + rule.consequent .mkString("[", ",", "]")
      + ", " + rule.confidence)
}
```

待模型训练结束后，即可在测试数据集上，使用该模型，对测试样本进行分类。

36.5 实验结果

FP-Growth 常用于“查看哪些商品经常被一起购买”“挖掘出哪些新闻广泛被用户浏览过”等场景，当需要使用 FP-Growth 算法找出数据项中最频繁项集时，首先便是构造 FP-tree，其结果是挖掘出“频繁项集”，如图 36-5 和图 36-6 所示。

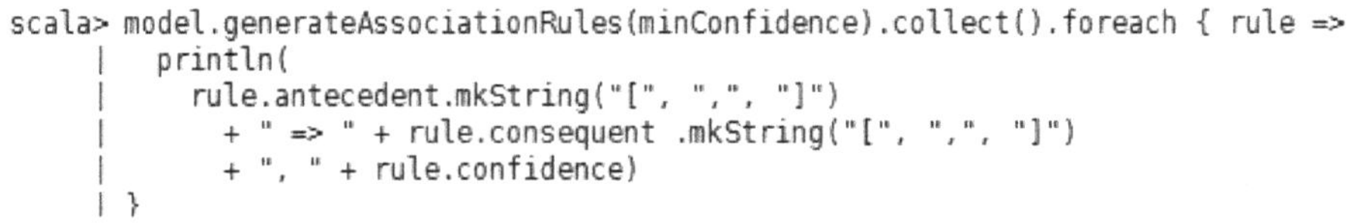

```
scala> model.generateAssociationRules(minConfidence).collect().foreach { rule =>
     |   println(
     |     rule.antecedent.mkString("[", ",", "]")
     |       + " => " + rule.consequent .mkString("[", ",", "]")
     |       + ", " + rule.confidence)
     | }
```

图 36-5 实验结果（一）

```
[q,y,x] => [z], 1.0
[t,z] => [x], 1.0
[t,z] => [y], 1.0
[y,s,x,z] => [t], 1.0
[q,y,t,x] => [z], 1.0
[t] => [y], 1.0
[t] => [x], 1.0
[t] => [z], 1.0
[q,t,z] => [x], 1.0
[q,t,z] => [y], 1.0
[p] => [r], 1.0
[p] => [z], 1.0
[y,t,x] => [z], 1.0
[y,s,t] => [x], 1.0
[y,s,t] => [z], 1.0
[y,z] => [t], 1.0
[y,z] => [x], 1.0
[q,y] => [t], 1.0
[q,y] => [x], 1.0
[q,y] => [z], 1.0
[y,s,t,z] => [x], 1.0
[q,t] => [x], 1.0
[q,t] => [z], 1.0
[q,t] => [y], 1.0
[q,y,t] => [x], 1.0
[q,y,t] => [z], 1.0
[q] => [t], 1.0
[q] => [y], 1.0
[q] => [x], 1.0
[q] => [z], 1.0
[q,t,x] => [z], 1.0
[q,t,x] => [y], 1.0
[s,t,x,z] => [y], 1.0
[q,y,t,z] => [x], 1.0
[p,r] => [z], 1.0
[r,z] => [p], 1.0
[y,s,t,x] => [z], 1.0
[t,x] => [y], 1.0
[t,x] => [z], 1.0
```

图 36-6　实验结果（二）

实验三十七　综合实战：车牌识别

37.1　实验目的

基于 MapReduce 思想，编写车牌识别程序，实现对江苏某两处监控图片中车牌的识别，完成对除江苏省车辆外的外省车辆的统计，并完成对两处监控中套牌车辆的识别，得出结果。

37.2　实验要求

要能理解 MapReduce 编程思想，编写 MapReduce 车牌识别程序，能够利用 Java 调用已封装好 C++的 so 动态库完成车牌的识别，然后利用 MapReduce 完成对车牌的统计和对套牌车的识别。最后将其执行并分析执行过程。

37.3　实验步骤

37.3.1　编写程序

对于外省车牌的统计，我们可以理解为 WordCount 程序，利用 WordCount 的思想，在 Reduce 阶段对“苏”车牌不进行输出即可。

对于套牌车辆的识别，Map 阶段与 WordCount 是一样的，在 Reduce 阶段我们将统计相同车牌出现的情况。

我们将编写两类，一类用于加载 C++的 so 动态库，另一类完成 MapReduce 算法。

SelectPic.java

```
import java.io.File;
import java.util.Vector;

public class SelectPic {
 private Vector<String> pics = null;

 public SelectPic() {
        pics = new Vector<>();
 }
```

```
    static {
        System.loadLibrary("Easyper");
    }

    public native String getPlate(String inPath);

    public void doPicNames(String inPath) {
        File file = new File(inPath);
        File[] listFiles = file.listFiles();
        for (File files : listFiles) {
            if (!files.isDirectory()) {
                String name = files.getName();
                String plate = getPlate(inPath + "/" + name);
                if (plate.length() > 3)
                    plate.substring(3);
            }
        }
    }

    public static void main(String[] args) {
        SelectPic sp = new SelectPic();
        sp.doPicNames("image");
    }

}
```

PlateRecog.java

```
import java.io.IOException;
import java.util.Vector;

import org.apache.hadoop.conf.Configuration;
import org.apache.hadoop.fs.FileSystem;
import org.apache.hadoop.fs.Path;
import org.apache.hadoop.io.IntWritable;
import org.apache.hadoop.io.LongWritable;
import org.apache.hadoop.io.NullWritable;
import org.apache.hadoop.io.Text;
import org.apache.hadoop.io.WritableComparable;
import org.apache.hadoop.io.WritableComparator;
import org.apache.hadoop.mapreduce.Job;
import org.apache.hadoop.mapreduce.Mapper;
import org.apache.hadoop.mapreduce.Reducer;
import org.apache.hadoop.mapreduce.lib.input.FileInputFormat;
import org.apache.hadoop.mapreduce.lib.output.FileOutputFormat;
import org.apache.hadoop.util.GenericOptionsParser;
```

```
public class PlateRecog {

    public static class PlateRecogMapper_coll extends Mapper<LongWritable, Text, Text, IntWritable> {

        private SelectPic sp = null;
        private FileSystem HDFS = null;

        @Override
        protected void setup(Mapper<LongWritable, Text, Text, IntWritable>.Context context)
                throws IOException, InterruptedException {
            Configuration conf = context.getConfiguration();
            sp = new SelectPic();
            HDFS = FileSystem.get(conf);
        }

        @Override
        protected void map(LongWritable key, Text value, Mapper<LongWritable, Text, Text, Int
Writable>.Context context)
                throws IOException, InterruptedException {
            String[] splits = value.toString().split(" ");
            String ds = splits[0].substring(0, splits[0].length() - 4) + "_t.jpg";
            Path src = new Path(splits[0]);
            Path dst = new Path(ds);
            HDFS.copyToLocalFile(src, dst);
            String plate = sp.getPlate(ds);
            plate = plate.substring(3);
            context.write(new Text(plate.substring(0, 1)), new IntWritable(1));
        }
    }

    // 统计非江苏的车牌的数量
    public static class PlateRecogReducer_coll extends Reducer<Text, IntWritable, Text, IntWritable> {
        @Override
        protected void reduce(Text k2, Iterable<IntWritable> v2s,
                Reducer<Text, IntWritable, Text, IntWritable>.Context context)
                throws IOException, InterruptedException {
            if (!k2.toString().equals("苏")) {
                int sum = 0;
                for (IntWritable v2 : v2s) {
                    sum += v2.get();
                }
                context.write(new Text(k2.toString() + " 车牌的数量："), new IntWritable(sum));
            }
```

```
        }
    }

    public static class PlateRecogMapper_find extends Mapper<LongWritable, Text, Text, Text> {

        private SelectPic sp = null;
        private FileSystem HDFS = null;

        @Override
        protected void setup(Mapper<LongWritable, Text, Text, Text>.Context context)
                throws IOException, InterruptedException {
            Configuration conf = context.getConfiguration();
            sp = new SelectPic();
            HDFS = FileSystem.get(conf);
        }

        @Override
        protected void map(LongWritable key, Text value, Mapper<LongWritable, Text, Text, Text>.
Context context)
                throws IOException, InterruptedException {
            String[] splits = value.toString().split(" ");
            String ds = splits[0].substring(0, splits[0].length() - 4) + "_t.jpg";
            Path src = new Path(splits[0]);
            Path dst = new Path(ds);
            HDFS.copyToLocalFile(src, dst);
            String plate = sp.getPlate(ds);
            plate = plate.substring(3);
            String plateInfo = plate + '_' + splits[1];

            // 皖A12345_1 45
            context.write(new Text(plateInfo), new Text(dst.getName().substring(0, dst.getName().length
() - 6)));
        }
    }

    public static class PlateRecogReducer_find extends Reducer<Text, Text, Text, NullWritable> {

        private Vector<String> samePlates = null;

        @Override
        protected void setup(Reducer<Text, Text, Text, NullWritable>.Context context)
                throws IOException, InterruptedException {
            samePlates = new Vector<>();
        }
```

```
        @Override
        protected void reduce(Text k2, Iterable<Text> v2s, Reducer<Text, Text, Text, NullWritable>.
Context context)
                    throws IOException, InterruptedException {
            int sum = 0;
            String[] splits = k2.toString().split("_");
            samePlates.clear();
            for (Text v2 : v2s) {
                ++sum;
                samePlates.add(v2.toString());
            }
            if (sum > 1) {
                String sameplate = "";
                for (int i = 0; i < samePlates.size() - 1; ++i) {
                        meplate += samePlates.get(i) + " 和 ";
                }
                sameplate += samePlates.get(samePlates.size() - 1);
                context.write(new Text("存在套牌的车辆的车牌号为 :" + splits[0] + " 车牌编号为：
" + sameplate), NullWritable.get());
            }
        }
    }

    public static class findGroupComparator extends WritableComparator {

        protected findGroupComparator() {
            super(Text.class, true);// 注册 comparator
        }

        @Override
        public int compare(WritableComparable a, WritableComparable b) {
            // TODO Auto-generated method stub
            Text ti1 = (Text) a;
            Text ti2 = (Text) b;
            return ti1.toString().substring(0, 7).compareTo(ti2.toString().substring(0, 7));
        }
    }

    public static void main(String[] args) {

        Configuration conf = new Configuration();
        try {
            GenericOptionsParser goparser = new GenericOptionsParser(conf, args);
```

```
            String otherargs[] = goparser.getRemainingArgs();

            Job job_coll = Job.getInstance(conf);
            job_coll.setJarByClass(PlateRecog.class);

            job_coll.setMapperClass(PlateRecogMapper_coll.class);
            job_coll.setReducerClass(PlateRecogReducer_coll.class);

            job_coll.setMapOutputKeyClass(Text.class);
            job_coll.setMapOutputValueClass(IntWritable.class);

            FileInputFormat.addInputPath(job_coll, new Path(otherargs[0]));
            FileOutputFormat.setOutputPath(job_coll, new Path(otherargs[1]));
            if ((job_coll.waitForCompletion(true) ? 1 : 0) == 1) {
                Job job_find = Job.getInstance(conf);
                job_find.setJarByClass(PlateRecog.class);

                job_find.setMapperClass(PlateRecogMapper_find.class);
                job_find.setReducerClass(PlateRecogReducer_find.class);

                job_find.setMapOutputKeyClass(Text.class);
                job_find.setMapOutputValueClass(Text.class);
                job_find.setOutputKeyClass(Text.class);
                job_find.setOutputValueClass(NullWritable.class);

                // job_find.setPartitionerClass(findParitioner.class);
                job_find.setGroupingComparatorClass(findGroupComparator.class);

                FileInputFormat.addInputPath(job_find, new Path(otherargs[0]));
                FileOutputFormat.setOutputPath(job_find, new Path(otherargs[2]));
                job_find.waitForCompletion(true);
            }
        } catch (IOException e) {
            // TODO Auto-generated catch block
            e.printStackTrace();
        } catch (ClassNotFoundException e) {
            // TODO Auto-generated catch block
            e.printStackTrace();
        } catch (InterruptedException e) {
            // TODO Auto-generated catch block
            e.printStackTrace();
        }
    }
}
```

37.3.2 环境准备

将实验数据上传到 HDFS 上，包括两部分，一部分是车牌照片和车牌对应的 plate.txt 文件，另一部分是所有车牌的照片。

在 client 机器上在 Hadoop 的 HDFS 中创建两个目录。

```
[root@client hadoop]# bin/hadoop  fs  -mkdir  -p  /user/mapreduce/platerecog/in
[root@client hadoop]# bin/hadoop  fs  -mkdir  /user/mapreduce/platerecog/images
```

上传数据。

```
[root@client hadoop]# bin/hadoop  fs  -put  /root/data/37/images/*  /user/mapreduce/platerecog/images
[root@client hadoop]# bin/hadoop  fs  -put  /root/data/37/in/plate.txt  /user/mapreduce/platerecog/in
```

在 master 机器上，将动态库所需要的配置文件复制至 Hadoop 根目录下。

```
[root@master hadoop]# \cp  -r  ../plate/use/etc/  ./
[root@master hadoop]# \cp  -r  ../plate/use/resources/  ./
[root@master hadoop]# cp  -r  ../plate/use/libEasyper.so  ./lib
```

然后，将动态库和配置文件复制至 slave1，slave2，slave3 上。

首先创建 machines 文件写入几台机器的主机名。

```
[root@ master hadoop]# vim machines
slave1
slave2
slave3
client
```

然后利用 for 进行远程复制。

```
[root@master hadoop]# for x in 'cat machines';do scp -r ../plate/use/etc/  $x:/usr/cstor/hadoop/;done;
[root@master hadoop]# for x in 'cat machines';do scp -r ./resources/ $x:/usr/cstor/hadoop/;done;
[root@master hadoop]# for x in 'cat machines';do scp -r ../plate/use/libEasyper.so  \
>  $x:/usr/cstor/hadoop/lib;done;
```

37.3.3 打包提交

使用 Eclipse 开发工具将该代码打包，选择主类为 PlateRecog。假如打包后的文件名为 PlateRecog.jar，主类 PlateRecog 位于默认包下，则可使用如下命令向 Hadoop 集群提交本应用。

```
[root@client hadoop]# bin/hadoop  jar  PlateRecog.jar  -files  \
> ../plate/use/libEasyper.so,../plate/use/etc/,../plate/use/resources/  \
> /user/mapreduce/platerecog/in/plate.txt  /user/mapreduce/platerecog/count/  \
> /user/mapreduce/platerecog/samePlate/
```

其中“hadoop”为命令，“jar”为命令参数，后面紧跟打的包，“-files”是附加文件指令，“../plate/use/libEasyper.so,../plate/use/etc/,../plate/use/resources/”是同时提交到集群上的配置文件和动态库。“/user/mapreduce/platerecog/in/plate.txt”为输入文件在 HDFS 中的位置，“/user/mapreduce/platerecog/count/”和 “/user/mapreduce/platerecog/samePlate/”为输出文件在 HDFS 中的位置，分别对应的是统计车牌的数量的输出文件 count，套牌车识别的结果文件 samePlate。

37.4　实验结果

37.4.1　输入数据

输入数据格式如下：plate.txt (空格（''）分割)（数据放在/root/data/37/in 目录下）。

```
/user/mapreduce/platerecog/images/99.jpg 2
/user/mapreduce/platerecog/images/100.jpg 2
/user/mapreduce/platerecog/images/101.jpg 2
/user/mapreduce/platerecog/images/102.jpg 2
/user/mapreduce/platerecog/images/103.jpg 2
/user/mapreduce/platerecog/images/104.jpg 2
/user/mapreduce/platerecog/images/105.jpg 2
/user/mapreduce/platerecog/images/106.jpg 2
/user/mapreduce/platerecog/images/107.jpg 2
......
```

37.4.2　执行结果

在 client 上执行对 HDFS 上的文件/user/mapreduce/platerecog/count/part-r-00000 内容查看的操作，查看外省车辆的统计情况。

```
[root@client hadoop]# bin/hadoop fs -cat  /user/mapreduce/platerecog/count/p*
```

外省车辆统计情况如图 37-1 所示。

```
[root@client hadoop]# bin/hadoop fs -cat  /user/mapreduce/platerecog/count/p*
16/12/15 01:03:16 WARN util.NativeCodeLoader: Unable to load native-hadoop library
e applicable
京 车牌的数量： 2
冀 车牌的数量： 1
吉 车牌的数量： 2
川 车牌的数量： 3
晋 车牌的数量： 4
沪 车牌的数量： 2
津 车牌的数量： 5
浙 车牌的数量： 21
渝 车牌的数量： 1
湘 车牌的数量： 1
甘 车牌的数量： 1
皖 车牌的数量： 17
粤 车牌的数量： 1
蒙 车牌的数量： 1
豫 车牌的数量： 1
鄂 车牌的数量： 3
青 车牌的数量： 1
鲁 车牌的数量： 2
```

图 37-1　查看外省车辆统计情况

在 client 上执行对 HDFS 上的文件/user/mapreduce/platerecog/samePlate/part-r-00000 内容查看的操作，查看两处监控的套牌车辆的情况。

```
[root@client hadoop]# bin/hadoop fs -cat  /user/mapreduce/platerecog/samePlate/p*
```

套牌情况如图 37-2 所示。

```
[root@client hadoop]# bin/hadoop fs -cat  /user/mapreduce/platerecog/samePlate/p*
16/12/15 01:03:26 WARN util.NativeCodeLoader: Unable to load native-hadoop library
e applicable
存在套牌的车辆的车牌号为 :浙F2Q001 车牌编号为:  51 和 118
```

图 37-2　查看套牌情况

我们将两个编号的车牌图片拿出来查看一下，如图 37-3 所示。

图 37-3　查看两个编号的车牌图片

实验三十八　综合实战：搜索引擎

38.1　实验目的

1. 利用大数据实验环境完成一个真实的项目；
2. 结合多个大数据的组件练习如何在实际项目中使用大数据的组件；
3. 能使用分布式的思想对数据进行清洗、处理；
4. 考虑大数据环境，对数据的处理方法进行优化；
5. 利用大数据的思路，了解通用的搜索引擎技术流程。

38.2　实验要求

在实验结束的时候能达到实验结果所展示的要求，能从 HBase 中正确查询到所需要的数据，并且结果能按照相关性大小排序。

能明白 HBase 建立索引的理由，并能理解 HBase 建立索引的方法，能正确使用索引，利用索引在 HBase 中作用。

能理解在扒取互联网数据时候的流程，在实验完成之后有兴趣可以对其他网站运用同样的方法对数据进行扒取。

能完整的将整个实验流程理解。了解大数据环境下利用 HBase 组件的一些功能。并能比较利用 NoSQL 的数据库和关系型数据库的区别。

38.3　实验步骤

我们实验的内容主要是先从淘宝的网页中扒取部分商品的信息，然后从这些信息中提取我们所需要的几项数据，将其存储在 HBase 中。然后我们可以通过关键词搜索从 HBase 中将这些数据根据我们的关键词提取出来。首先，我们需要在准备好的 HBase 中创建数据表 goodinfo 和索引表 indexinfo。

建表语句：

```
hbase(main):001:0>   create 'goodinfo', 'info'
hbase(main):002:0>   create 'indexinfo', 'info'
```

38.3.1　新建 Java 项目

因为 HBase 只能使用 rowKey 从 HBase 中读取数据。而我们需要使用关键词从

HBase 读取数据，所以我们需要在商品的标题中提取关键词，然后对关键词和 URL 建立索引。这样我们就可以通过商品的标题对分词过后的关键词进行搜索查询到商品的 URL，然后通过 URL 取得商品的信息。

新建 Createindex 项目，主要分为几个步骤，新建 Hadoop 任务的 Map 类和 Reduce 类，新建任务类。

项目目录结构如图 38-1 所示。

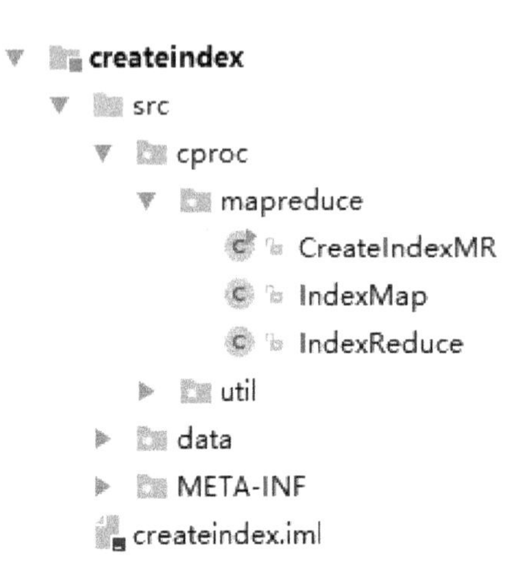

图 38-1　项目目录结构（一）

项目所需的 jar 可在实验的/root/data/38/jar 使用 ftp 下载到本地即可。

新建 IndexMap 类，Map 类的代码如下。

```
package cproc.mapreduce;

import java.io.IOException;
import java.util.ArrayList;
import java.util.HashMap;
import java.util.List;
import java.util.Map;

import org.apache.hadoop.hbase.client.Result;
import org.apache.hadoop.hbase.io.ImmutableBytesWritable;
import org.apache.hadoop.hbase.mapreduce.TableMapper;
import org.apache.hadoop.hbase.util.Bytes;
import org.apache.hadoop.io.Text;
import org.apache.hadoop.mapreduce.Mapper;

import cproc.util.BuildSegmenter;
import edu.stanford.nlp.ie.crf.CRFClassifier;
import edu.stanford.nlp.ling.CoreLabel;
```

```
public class IndexMap extends TableMapper<Text, Text> {

    CRFClassifier<CoreLabel> segmenter = null;

    @Override
    protected void setup(
            Mapper<ImmutableBytesWritable, Result, Text, Text>.Context context)
            throws IOException, InterruptedException {
        super.setup(context);

        // 初始化分词器 —— 加载速度慢，该步骤仅操作一次，耗时约 10s
        if(segmenter==null){
            BuildSegmenter bs = new BuildSegmenter();
            segmenter = bs.BuildSegmenter();
        }
    }

    @Override
    public void map(ImmutableBytesWritable key, Result value, Context context)
            throws IOException, InterruptedException {

        String URL = Bytes.toString(key.get());
        Map<String, Integer> wordmap = null;

        String title = "";
        byte[] bytes = value.getValue(Bytes.toBytes("info"), Bytes.toBytes("itemTitle"));
        if(bytes!=null && bytes.length>0){
            title = Bytes.toString(bytes);
        }

        //标题分词
        List<String> keyWordSeg = new ArrayList<String>();
        keyWordSeg = segmenter.segmentString(title);

        //每个 URL 的标题内单词合并
        wordmap = new HashMap<String, Integer>();
        for(String word : keyWordSeg){
            if(wordmap.containsKey(word)){
                wordmap.put(word, (Integer)wordmap.get(word)+1);
            }else{
                wordmap.put(word, 1);
            }
        }
```

```
        for(String word : wordmap.keySet()){
            context.write(new Text(word),new Text(wordmap.get(word)+"##"+URL));
        }

    }
}
```

新建 IndexReduce 类，Reduce 类的代码如下。

```
package cproc.mapreduce;

import java.io.IOException;

import org.apache.hadoop.hbase.client.Put;
import org.apache.hadoop.hbase.io.ImmutableBytesWritable;
import org.apache.hadoop.hbase.mapreduce.TableReducer;
import org.apache.hadoop.hbase.util.Bytes;
import org.apache.hadoop.io.Text;

public class IndexReduce extends TableReducer<Text, Text, ImmutableBytesWritable>
{

    @Override
    protected void reduce(Text key, Iterable<Text> values, Context context) throws IOException,
InterruptedException
    {
        String str = "";

        for (Text value : values) {
            str += value.toString() + "####";
         }

        if(str!=null && str.length()>4)
        {
            str = str.substring(0, str.length()-4);
        }
        Put put = null;
        try{
            put = new Put(Bytes.toBytes(key.toString()));

        }catch (Exception e){
            System.out.println("key is :      "+ key.toString());
            System.out.println(e.getMessage());
        }
        if(put!=null) {
            // 列簇 ，列名, 值
```

```
            put.addColumn(Bytes.toBytes("info"), Bytes.toBytes("val"), Bytes.toBytes(str));

            context.write(new ImmutableBytesWritable(Bytes.toBytes(key.toString())), put);
        }

    }

}
```

新建 CreateIndexMR，新建 MapReduce 的主方法。

```
package cproc.mapreduce;

import org.apache.hadoop.conf.Configuration;
import org.apache.hadoop.filecache.DistributedCache;
import org.apache.hadoop.fs.Path;
import org.apache.hadoop.hbase.HBaseConfiguration;
import org.apache.hadoop.hbase.client.Scan;
import org.apache.hadoop.hbase.mapreduce.TableMapReduceUtil;
import org.apache.hadoop.hdfs.DistributedFileSystem;
import org.apache.hadoop.io.Text;
import org.apache.hadoop.mapreduce.Job;


public class CreateIndexMR {

    @SuppressWarnings("deprecation")
    public static void main(String[] args) throws Exception {

        //初始化 conf 文件
        Configuration conf = HBaseConfiguration.create();
        //HBase 配置
        conf.set("hbase.zookeeper.quorum", "slave1:2181,slave2:2181,slave3:2181");//你的 Zookeeper 的地址
        conf.set("zookeeper.znode.parent", "/hbase");

        //定义任务
        Job job = new Job(conf, "createIndex");
         job.setJarByClass(CreateIndexMR.class);


         Scan scan = new Scan();
          scan.setCaching(500);            // 默认是 1，应该设置得大一些
          scan.setCacheBlocks(false);  // 当 mr 或者全表查询的时候要设置成 false，查询热数据的的时
候可以使用，这样就在 HBase 的 jvm 中缓存当前记录

          //设置 map
```

```
        TableMapReduceUtil.initTableMapperJob(
                "goodinfo",              // input table
                scan,                    // Scan
                IndexMap.class,          // mapper class
                Text.class,              // mapper output key
                Text.class,              // mapper output value
                job);

        //设置 reduce (output table , reducer class , job)
         TableMapReduceUtil.initTableReducerJob("indexinfo", IndexReduce.class, job);

        // 运行 job
        System.exit(job.waitForCompletion(true) ? 0 : 1);

    }

}
```

新建 BuildSegmenter 类，新建分词工具类。

```
package cproc.util;

import java.util.Properties;

import edu.stanford.nlp.ie.crf.CRFClassifier;
import edu.stanford.nlp.ling.CoreLabel;

/**
 * 初始化分词器
 */

public class BuildSegmenter {
    public CRFClassifier<CoreLabel> BuildSegmenter() {
        /**
         * 初始化分词器
         * @return
         */
        //String basedir = System.getProperty("Sousuo", "data");
//      String basedir = BuildSegmenter.class.getClassLoader().getResource("data").getPath();
        String basedir = "/data";

        Properties props = new Properties();
        props.setProperty("sighanCorporaDict", basedir);
        props.setProperty("serDictionary", basedir + "/dict-chris6.ser.gz");

        props.setProperty("inputEncoding", "UTF-8");
```

```
        props.setProperty("sighanPostProcessing", "true");

        CRFClassifier<CoreLabel> segmenter = new CRFClassifier<>(props);
        System.out.println(basedir);
        segmenter.loadClassifierNoExceptions(basedir + "/ctb.gz", props);

        return segmenter;
    }
}
```

到此，CreateIndex 代码部分已经完成。Createindex 项目的作用是对 HBase 中的 goodinfo 集合数据建立索引表 indexinfo。

38.3.2 新建 JavaWeb 项目

该项目的主要作用是提供 Web 界面扒取淘宝的数据和搜索数据。该项目部署在本地服务器上，如图 38-2 所示。

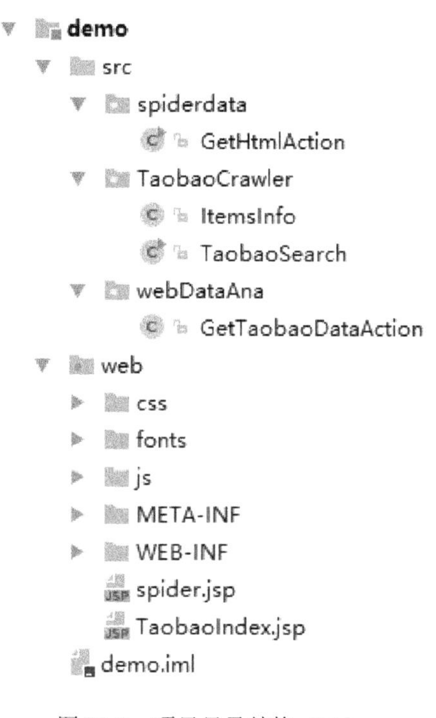

图 38-2 项目目录结构（二）

对于 Web 目录下的除 jsp 页面外的代码放在/root/data/38/web 目录下，请读者自行下载加入自己的目录下。

该项目中存在分词器的工具，该工具是必须使用 jdk1.8 版本才能正常运行，所以在创建项目之前需要安装 jdk1.8 到本地，该 JavaWeb 项目需要依赖于该 jdk。请自行修改，修改步骤如下。

单击 File→project Structure→选择 modules→单击新建的 JavaWeb 项目→单击如下位置。

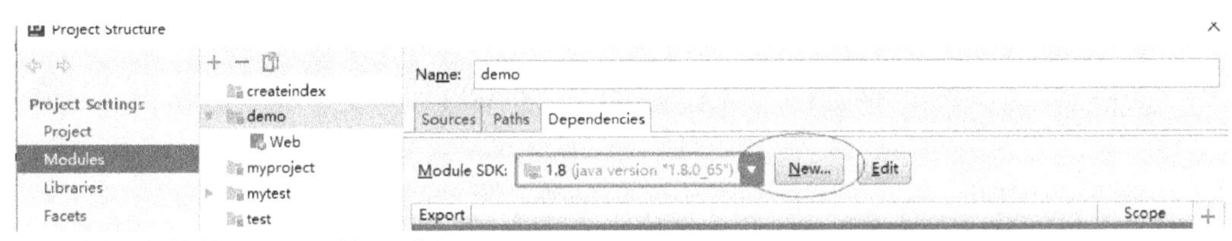

选择安装的 jdk1.8 的目录即可。

新建 spider.jsp 文件，作为页面扒取的 Web 页面。

```
<!-- you_shoubian -->
<%@page import="TaobaoCrawler.ItemsInfo"%>
<%@page import="Java.util.*"%>
<%@ page import="WebDataAna.GetTaobaoDataAction" %>
<%@ page language="Java" contentType="text/html; charset=UTF-8"
        pageEncoding="UTF-8"%>


<%
    String searchWord = "";
    String pageNumStr = "";
    String itemLen = "";

    String msg = "";
%>

<!DOCTYPE html PUBLIC "-//W3C//DTD HTML 4.01 Transitional//EN" "http://www.w3.org/TR/
html4/loose.dtd">
<html>
<head>
    <meta http-equiv="Content-Type" content="text/html; charset=UTF-8">
    <title>淘宝集合</title>
    <link rel="stylesheet" href="css/bootstrap.min.css">
    <link rel="stylesheet" href="css/myTaobaoCss.css">
    <script type="text/Javascript" src="js/jquery-1.11.3.min.js"></script>
    <script type="text/Javascript" src="js/bootstrap.min.js"></script>
</head>
<body>
<div class="container">
    <div style="height: 100px"></div>

    <div class="formStyle col-md-offset-1 col-md-10 ">
        <form class="" method="post" action="/getHtml">
            <div class="form-group col-md-7 mySearchMar">
```

```
                <input type="text" class="form-control " id="searchWords"
                        name="searchWords" placeholder="请输入搜索的内容,多个关键词用 ，隔开">
            </div>
            <div class="form-group col-md-2 myPageMar">
                <div class="input-group">
                    <input type="text" class="form-control" id="pageNum" name="pageNum"
                            placeholder="1">
                    <div class="input-group-addon">页</div>
                </div>
            </div>


            <button type="submit" class="btn btn-default col-md-1"
                    onclick="showMsg()">检索</button>
        </form>
    </div>

    <div class="tableStyle col-md-offset-1 col-md-10">
        <div id="msg"></div>
    </div>



    <%
        if (request.getAttribute("searchWord") != null) {
            searchWord = (String) request.getAttribute("searchWord");
            itemLen = (String) request.getAttribute("itemLen");
            String pageNum = (String) request.getAttribute("pageNumStr");

            msg = "共获取 <b>" + pageNum + "</b> 页,获取数据 <b>" + itemLen + "</b> 条,搜索词
为 <b>"+searchWord+"</b>";
            if(searchWord.equals("无搜索词，结果随机搜索")){
                searchWord="";
            }

        }
    %>


</div>

<script type="text/Javascript">
    $("#searchWord").val("<%=searchWord%>");
    $("#pageNum").val("<%=pageNumStr%>");
    $("#msg").html("<%=msg%>");
```

```
    function showMsg() {
        $("#msg").text("正在检索...");
    }
</script>
</body>
</html>
```

新建 TaobaoIndex.jsp 文件，作为 Taobao 数据展示 Web 页面。

```
<!-- you_shoubian -->
<%@page import="TaobaoCrawler.ItemsInfo"%>
<%@page import="java.util.*"%>
<%@ page import="WebDataAna.GetTaobaoDataAction" %>
<%@ page language="Java" contentType="text/html; charset=UTF-8"
    pageEncoding="UTF-8"%>


<%
    String searchWord = "";
    String pageNumStr = "";
    String msg = "";
%>

<!DOCTYPE html PUBLIC "-//W3C//DTD HTML 4.01 Transitional//EN" "http://www.w3.
org/TR/html4/loose.dtd">
<html>
<head>
<meta http-equiv="Content-Type" content="text/html; charset=UTF-8">
<title>淘宝集合</title>
<link rel="stylesheet" href="css/bootstrap.min.css">
<link rel="stylesheet" href="css/myTaobaoCss.css">
<script type="text/Javascript" src="js/jquery-1.11.3.min.js"></script>
<script type="text/Javascript" src="js/bootstrap.min.js"></script>
</head>
<body>
    <div class="container">
        <div style="height: 100px"></div>

        <div class="formStyle col-md-offset-1 col-md-10 ">
            <form class="" method="post" action="/getTaobaoData">
                <div class="form-group col-md-7 mySearchMar">
                    <input type="text" class="form-control " id="searchWord"
                        name="searchWord" placeholder="请输入搜索的内容">
                </div>
                <div class="form-group col-md-2 myPageMar">
```

```
            <div class="input-group">
                <input type="text" class="form-control" id="pageNum" name="pageNum"
                    placeholder="1">
                <div class="input-group-addon">项</div>
            </div>
        </div>


        <button type="submit" class="btn btn-default col-md-1"
            onclick="showMsg()">检索</button>
    </form>
</div>

<div class="tableStyle col-md-offset-1 col-md-10">
    <div id="msg"></div>
</div>

<div class="tableStyle col-md-offset-1 col-md-10">
    <table class="table table-striped">
        <thead>
            <tr>
                <th>#</th>
                <th>商品名</th>
                <th class="col-md-1">价格</th>
                <th>交易量</th>
                <th class="col-md-2">店铺</th>
                <th>所在地</th>
            </tr>
        </thead>
        <tbody>
            <%
                List<ItemsInfo> itemsInfoList = new ArrayList<ItemsInfo>();
                if (request.getAttribute("itemsInfoList") != null) {
                    itemsInfoList = (ArrayList<ItemsInfo>) request.getAttribute("itemsInfoList");
                    searchWord = (String) request.getAttribute("searchWord");
                    pageNumStr = (String) request.getAttribute("pageNumStr");
                    int pageNum = (Integer) request.getAttribute("pageNum");

                    msg = "共获取数据 <b>" + itemsInfoList.size() + "</b> 条,搜索词为
<b>"+searchWord+"</b>";
                    if("无搜索词，结果随机搜索".equals(searchWord)){
                        searchWord="";
                    }
```

```
                        int len = itemsInfoList.size();
                        for (int i = 0; i < len; i++) {
                %>
                <tr>
                    <td><%=i + 1%></td>
                    <td><a href="<%=itemsInfoList.get(i).getItemAddress()%>"
                        title="点击查看详情"><%=itemsInfoList.get(i).getItemTitle()%></a></td>
                    <td><%=itemsInfoList.get(i).getItemPrice()%></td>
                    <td><%=itemsInfoList.get(i).getItemDealCnt()%></td>
                    <td><%=itemsInfoList.get(i).getItemShop()%></td>
                    <td><%=itemsInfoList.get(i).getItemShopLocaltion()%></td>
                </tr>
                <%
                    }
                    }
                %>

            </table>
        </div>

    </div>

    <script type="text/Javascript">
        $("#searchWord").val("<%=searchWord%>");
        $("#pageNum").val("<%=pageNumStr%>");
        $("#msg").html("<%=msg%>");

        function showMsg() {
            $("#msg").text("正在检索...");
        }
    </script>
</body>
</html>
```

创建 servlet 类、GetHtmlAction 类，这个 servlet 访问/getHtml 的 http 请求。

```
package spiderdata;

import java.io.File;
import java.io.IOException;
import java.util.ArrayList;
import java.util.List;

import javax.servlet.ServletException;
import javax.servlet.annotation.WebServlet;
import javax.servlet.http.HttpServlet;
```

```
import javax.servlet.http.HttpServletRequest;
import javax.servlet.http.HttpServletResponse;

import org.apache.hadoop.conf.Configuration;
import org.apache.hadoop.hbase.HBaseConfiguration;
import org.apache.hadoop.hbase.TableName;
import org.apache.hadoop.hbase.client.Connection;
import org.apache.hadoop.hbase.client.ConnectionFactory;
import org.apache.hadoop.hbase.client.Put;
import org.apache.hadoop.hbase.client.Table;
import org.apache.hadoop.hbase.util.Bytes;
import org.jsoup.Jsoup;
import org.jsoup.nodes.Document;
import org.jsoup.nodes.Element;
import org.jsoup.select.Elements;

import TaobaoCrawler.ItemsInfo;
import edu.stanford.nlp.ie.crf.CRFClassifier;
import edu.stanford.nlp.ling.CoreLabel;

/**
 * Servlet implementation class GetTaobaoDataAction
 */
@WebServlet("/getHtml")
public class GetHtmlAction extends HttpServlet {
  private static final long serialVersionUID = 1L;
  static String jsFile = "D:\\casperJS\\UrlCatch.js";
  static String tempfile = "D:\\tempHTML\\temp.html";
  static List<ItemsInfo> itemsInfoList = new ArrayList<ItemsInfo>();
  CRFClassifier<CoreLabel> segmenter = null;

  //HBase
  private static final Configuration configuration = HBaseConfiguration.create();
  private static Connection connection;
  static {

    configuration.set("hbase.zookeeper.quorum", " slave1:2181,slave2:2181,slave3:2181");
    configuration.set("zookeeper.znode.parent", "/hbase");

    try {
      connection = ConnectionFactory.createConnection(configuration);
    } catch (IOException e) {
      e.printStackTrace();
    }
```

```
  }

  public GetHtmlAction() {
    super();
  }

  protected void doGet(HttpServletRequest request, HttpServletResponse response) throws ServletException,
IOException {
    doPost(request, response);
  }

  protected void doPost(HttpServletRequest request, HttpServletResponse response) throws ServletException,
IOException {
    // TODO Auto-generated method stub
    response.setContentType("text/html;charset=UTF-8");
    response.setHeader("content-type", "text/html;charset=utf-8");
    response.setCharacterEncoding("UTF-8");

    try {
      String searchWords = new String(
            request.getParameter("searchWords")
                .getBytes("iso-8859-1"), "utf-8");
      String[] listWords = searchWords.split(",");
      String pageNumStr = new String(request.getParameter("pageNum").getBytes("iso-8859-1"), "utf-8");
      int pageNum = 1;
      try {
        pageNum = Integer.parseInt(pageNumStr);

      } catch (NumberFormatException e) {
        System.out.println("页数不是整数，默认为 1");
        pageNum = 1;
      }

      if(listWords==null || listWords.length <= 0){
        System.out.println("无搜索词");
        return;
      }
      int all = saveGoods(listWords,pageNum);
      request.removeAttribute("searchWord");
      request.removeAttribute("itemLen");
      request.removeAttribute("pageNumStr");

      request.setAttribute("searchWord",searchWords.toString());
      request.setAttribute("itemLen",all+"");
```

```
        request.setAttribute("pageNumStr",pageNum+"");
        request.getRequestDispatcher("spider.jsp").forward(request, response);

    } catch (Exception e) {
        e.printStackTrace();
    }
}


public  int saveGoods(String[] ListWords, int pageNum) throws Exception
{
    int sum = 0;

    for(String searchWord : ListWords)
    {
        System.out.println(searchWord);
        boolean flag = false;
            //初始化每页商品个数
            int cnt = 1;
            //初始化从第 s 个商品开始爬取
            int s = 1;
            for (int i = 1; i <= pageNum; i++) {
                System.out.println("正在获取第" + i + "/" + pageNum + "页...");
                Document doc = getHTMLDoc(jsFile, searchWord, s);
                itemsInfoList.clear();
                getItemsInfo(doc);
                if (!flag) {
                    cnt = itemsInfoList.size();
                    flag = true;
                }
                s = i * cnt;

                //保存 HBase
                TableName tbname = TableName.valueOf("goodinfo");
                Table table = connection.getTable(tbname);
                //商品对象集合
                List<Put> putList = new ArrayList<Put>();
                Put put = null;
                for(ItemsInfo info : itemsInfoList)
                {

                    sum ++;
                    //商品信息
                    put = new Put(info.getItemAddress().getBytes());
```

```
                put.addColumn(Bytes.toBytes("info"),
                    Bytes.toBytes("itemTitle"),
                    Bytes.toBytes(info.getItemTitle()));
                put.addColumn(Bytes.toBytes("info"),
                    Bytes.toBytes("itemShop"),
                    Bytes.toBytes(info.getItemShop()));
                put.addColumn(Bytes.toBytes("info"),
                    Bytes.toBytes("itemShopLocaltion"),
                    Bytes.toBytes(info.getItemShopLocaltion()));
                put.addColumn(Bytes.toBytes("info"),
                    Bytes.toBytes("itemPrice"),
                    Bytes.toBytes(info.getItemPrice()));
                put.addColumn(Bytes.toBytes("info"),
                    Bytes.toBytes("itemDealCnt"),
                    Bytes.toBytes(info.getItemDealCnt()));
                putList.add(put);
            }
            table.put(putList);
            table.close();
        }
            }
    return sum;
}


/* 得到 html 页面 */
public static Document getHTMLDoc(String jsFile, String searchWord, int s) {
    String tempfile = "D:\\tempHTML\\temp.html";

    try {
        String command = "cmd /c casperjs "+jsFile + " --url=https://s.taobao.com/search?q="
                + searchWord + "^&s=" + s + " --tempfile=d:\\tempHTML\\temp.html";

        System.out.println(command);
        Process process = Runtime.getRuntime().exec(command);
        process.waitFor();

        // 获取 html...
        File htmlfile = new File(tempfile);
        Document doc = Jsoup.parse(htmlfile, "UTF-8");
        return doc;

    } catch (Exception e) {
        e.printStackTrace();
```

```
        return null;
    }
}

/* 获取宝贝信息 */
public static void getItemsInfo(Document doc) {

    Elements elements = doc.getElementsByClass("ctx-box");

    for (Element el : elements) {
        ItemsInfo item = new ItemsInfo();

        String ItemTitle = el.getElementsByClass("title").first().text();
        String ItemShop = el.getElementsByClass("shop").first().text();
        String ItemShopLocaltion = el.getElementsByClass("location").first().text();
        String ItemPrice = el.getElementsByClass("price").first().text();
        String ItemDealCnt = el.getElementsByClass("deal-cnt").first().text();

        Element alinkElement = el.getElementsByClass("title").first();
        String ItemAddress = alinkElement.getElementsByTag("a").attr("href");
        if (ItemAddress.startsWith("//")) {
            ItemAddress = "https:" + ItemAddress;
        }

        item.setItemTitle(ItemTitle);
        item.setItemShop(ItemShop);
        item.setItemShopLocaltion(ItemShopLocaltion);
        item.setItemPrice(ItemPrice);
        item.setItemDealCnt(ItemDealCnt);
        item.setItemAddress(ItemAddress);

        itemsInfoList.add(item);
    }
}


public static void main(String[] args) throws Exception {
    String ss = "手机壳,硬盘";
    String[] ListWords = ss.split(",");
    int pageNum = 2;
    GetHtmlAction g = new GetHtmlAction();
    g.saveGoods(ListWords, pageNum);
}
```

```
}
```

创建 ItemInfo 类，作为从数据库读取数据的 model 类。

```
package TaobaoCrawler;
/**
 * author:you_shoubian time:2016-1-19
 */
public class ItemsInfo {

    String itemTitle;              //商品标题(名称)
    String itemShop;               //商品店铺名
    String itemShopLocaltion;  //商品店铺所在地
    String itemPrice;              //商品价格
    String itemDealCnt;                //商品交易量
    String itemAddress;                //商品详细地址



    public String getItemTitle() {
        return itemTitle;
    }
    public void setItemTitle(String itemTitle) {
        this.itemTitle = itemTitle;
    }
    public String getItemShop() {
        return itemShop;
    }
    public void setItemShop(String itemShop) {
        this.itemShop = itemShop;
    }
    public String getItemShopLocaltion() {
        return itemShopLocaltion;
    }
    public void setItemShopLocaltion(String itemShopLocaltion) {
        this.itemShopLocaltion = itemShopLocaltion;
    }
    public String getItemPrice() {
        return itemPrice;
    }
    public void setItemPrice(String itemPrice) {
        this.itemPrice = itemPrice;
    }
    public String getItemDealCnt() {
```

```
        return itemDealCnt;
    }
    public void setItemDealCnt(String itemDealCnt) {
        this.itemDealCnt = itemDealCnt;
    }
    public String getItemAddress() {
        return itemAddress;
    }
    public void setItemAddress(String itemAddress) {
        this.itemAddress = itemAddress;
    }


}
```

创建 Taobaosearch 类，用于从 HBase 中读取淘宝数据。

```
package TaobaoCrawler;

import org.apache.hadoop.conf.Configuration;
import org.apache.hadoop.HBase.HBaseConfiguration;
import org.apache.hadoop.HBase.TableName;
import org.apache.hadoop.HBase.client.*;
import org.apache.hadoop.HBase.util.Bytes;
import org.jsoup.Jsoup;
import org.jsoup.nodes.*;
import org.jsoup.select.Elements;

import Java.io.File;
import Java.io.IOException;
import Java.util.*;

/**
 * author:you_shoubian time:2016-1-19
 */
public class TaobaoSearch {
    private static final Configuration configuration = HBaseConfiguration.create();
    private static Connection connection;
    static {

        configuration.set("hbase.zookeeper.quorum", "slave1:2181,slave2:2181,slave3:2181");
        configuration.set("zookeeper.znode.parent", "/hbase");

        try {
            connection = ConnectionFactory.createConnection(configuration);
```

```
        } catch (IOException e) {
            e.printStackTrace();
        }
    }
    public static void main(String args[]) {
        try{
            getItemsInfoList("包邮",2);
        }catch (Exception e){
            e.printStackTrace();
        }


    }

    public static List<ItemsInfo> getItemsInfoList(String searchWord, int pNum) throws IOException {
        List<ItemsInfo> itemsInfoList = new ArrayList<ItemsInfo>();
        TableName tbname = TableName.valueOf("indexinfo");
        Table table = null;
        Result result = null;
        table = connection.getTable(tbname);
        result = table.get(new Get(Bytes.toBytes(searchWord)));
        byte[] byteURLs = result.getValue(Bytes.toBytes("info"), Bytes.toBytes("val"));
        String URLs = "";
        if(byteURLs!=null && byteURLs.length>0)
        {
            URLs = new String(byteURLs);
        }

        //带个数的 URL(次数##URL),每个 URL 中间用####隔开
        String[] comURLarr = null;
        if(URLs != null && URLs.length()>0)
        {
            comURLarr = URLs.split("####");
        }

        Map<String,Integer> URLmap = new HashMap<>();
        for(String URLandnum : comURLarr){
            String[] tim = URLandnum.split("##");
            URLmap.put(tim[1],Integer.parseInt(tim[0]));
        }
        //查询商品信息表
        if(comURLarr != null && comURLarr.length>0)
        {
```

```
        TableName tbname2 = TableName.valueOf("goodinfo");
        Table table2 = null;
        table2 = connection.getTable(tbname2);
        List<Get> list = new ArrayList<Get>();

        Get get = null;
        for(String comURL : comURLarr)
        {
            String[] arr = comURL.split("##");
            if(arr!=null && arr.length>1){
                get = new Get(Bytes.toBytes(arr[1]));
                list.add(get);
            }
        }

        Result[] results = table2.get(list);

        for(Result re : results)
        {
            System.out.println(Bytes.toString(re.getRow()) + ": " + Bytes.toString (re.getValue
(Bytes.toBytes("info"), Bytes.toBytes("itemTitle"))));
        }
        List<Object[]> itemsList = new ArrayList<>();

        for(Result res : results){
            ItemsInfo item = new ItemsInfo();
            Object[] item2 = new Object[2];
            item.setItemAddress(Bytes.toString(res.getRow()));

            item.setItemTitle(Bytes.toString(res.getValue(Bytes.toBytes("info"),Bytes.toBytes("itemTitle"))));
            item.setItemShop(Bytes.toString(res.getValue(Bytes.toBytes("info"),Bytes.toBytes("itemShop"))));
item.setItemShopLocaltion(Bytes.toString(res.getValue(Bytes.toBytes("info"),Bytes.toBytes
("itemShopLocaltion"))));
            item.setItemPrice(Bytes.toString(res.getValue(Bytes.toBytes("info"),Bytes.toBytes("itemPrice"))));
item.setItemDealCnt(Bytes.toString(res.getValue(Bytes.toBytes("info"),Bytes.toBytes ("itemDealCnt"))));
            item2[0] = item;
            item2[1] = URLmap.get(item.getItemAddress());
            itemsList.add(item2);
        }
        Collections.sort(itemsList, new Comparator<Object[]>() {
            @Override
            public int compare(Object[] o1, Object[] o2) {
```

```
                if(o1.length != 2 || o2.length != 2 ){
                    return 0;
                }
                if((int)o1[1] < (int)o2[1]){
                    return 1;
                }else{
                    return -1;
                }
            }
        });


        for(Object[] objarr : itemsList.subList(0,Math.min(pNum,itemsList.size()))){
            System.out.println(objarr[1]);
            itemsInfoList.add((ItemsInfo) objarr[0]);
        }
    }

    return itemsInfoList;
  }

}
```

创建 GetTaobaoDataAction servlet 类，接收/getTaobaoData 的 http 请求。

```
package webDataAna;

import TaobaoCrawler.ItemsInfo;
import TaobaoCrawler.TaobaoSearch;

import javax.servlet.ServletException;
import javax.servlet.annotation.WebServlet;
import javax.servlet.http.HttpServlet;
import javax.servlet.http.HttpServletRequest;
import javax.servlet.http.HttpServletResponse;
import java.io.IOException;
import java.util.ArrayList;
import java.util.List;

/**
 * Servlet implementation class GetTaobaoDataAction
 */
```

```
@WebServlet("/getTaobaoData")
public class GetTaobaoDataAction extends HttpServlet {
    private static final long serialVersionUID = 1L;

     /**
      * @see HttpServlet#HttpServlet()
      */
     public GetTaobaoDataAction() {
          super();
          // TODO Auto-generated constructor stub
     }

    /**
     * @see HttpServlet#doGet(HttpServletRequest request, HttpServletResponse response)
     */
    protected void doGet(HttpServletRequest request, HttpServletResponse response) throws ServletException,
IOException {
        // TODO Auto-generated method stub
        doPost(request, response);
    }

    /**
     * @see HttpServlet#doPost(HttpServletRequest request, HttpServletResponse response)
     */
    protected void doPost(HttpServletRequest request, HttpServletResponse response) throws ServletException,
IOException {
        // TODO Auto-generated method stub
        response.setContentType("text/html;charset=UTF-8");
        response.setHeader("content-type", "text/html;charset=utf-8");
        response.setCharacterEncoding("UTF-8");

        try {
            String searchWord = new String(request.getParameter("searchWord").getBytes("iso-8859-1"), "utf-8");
            String pageNumStr = new String(request.getParameter("pageNum").getBytes("iso-8859-1"), "utf-8");
             int pageNum = 1;
             try {
                 pageNum = Integer.parseInt(pageNumStr);

             } catch (NumberFormatException e) {
                 // TODO: handle exception
                 System.out.println("页数不是整数，默认为 1");
                 pageNum = 1;
```

```
        }

        if(searchWord.isEmpty()){
            searchWord = "无搜索词，结果随机搜索";
        }

        //System.out.println("/"+searchWord+"/ " + pageNum+" "+searchWord.isEmpty());
        List<ItemsInfo> itemsInfoList = null;
        try {
               itemsInfoList = TaobaoSearch.getItemsInfoList(searchWord, pageNum);
        }catch (Exception e){
            e.printStackTrace();
            itemsInfoList = new ArrayList<>();
        }
        int itemsLen = itemsInfoList.size();
        request.removeAttribute("itemsInfoList");
        request.removeAttribute("searchWord");
        request.removeAttribute("pageNumStr");
        request.removeAttribute("pageNum");


        request.setAttribute("itemsInfoList", itemsInfoList);
        request.setAttribute("searchWord", searchWord);
        request.setAttribute("pageNumStr", pageNumStr);
        request.setAttribute("pageNum", pageNum);

        request.getRequestDispatcher("TaobaoIndex.jsp").forward(request, response);


    } catch (Exception e) {
        // TODO: handle exception
    }
  }

}
```

这个项目的全部代码已经部署完毕，你只需要将这 JavaWeb 项目部署到 Tomcat 上就可以运行了，但是在运行之前还需要配置一下你本机的 Hosts 文件，用于连接 ZooKeeper。将你的集群中的机器的 hostname 和 IP 添加到你的本地 host 文件中。

38.3.3 网页扒取

因为淘宝在自己的网页上采用了反爬虫技术，用 Java 代码不容易扒取到正确的数据，所以我们需要采用一些外部的插件去获取淘宝网页，为了使用这些外部的插件我们需要提前准备做一些准备。

我们需要从服务器上使用 ftp 下载 phantomjs-1.9.8-windows，n1k0-casperjs-1.1-beta3-0-g4f105a9，然后将它们添加到本地的 PATH 中，对 phantomjs-1.9.8-windows，只需要将 phantomjs-1.9.8-windows/目录下的 phantomjs-1.9.8-windows 加入到本地 PATH 中，n1k0-casperjs-1.1-beta3-0-g4f105a9 要将 n1k0-casperjs-4f105a9\bin 添加到本地 PATH 中。再从服务器上下载 casperjs 文件夹，把这个目录放到 D 盘的根目录下，在 D 盘的根目录建立一个 tempHTML 目录，最后在里面添加 temp.html 文件。

至此，准备工作就做完了。

接下来我们打开 Tomcat，访问 spider.jsp 页面，如图 38-3 所示。

图 38-3　spider.jsp 页面

输入你想要扒取的关键词，可以同时扒取多个关键词，用逗号隔开。输入扒取的页数。稍等一会就可以看到扒取的结果（因为淘宝对同一个 IP 的单位时间淘宝页面的次数，所以当扒取的数量比较多的时候会比较慢。）

注：如果实验室无法连接外网，可以下载/root/data/38/data.csv 文件，然后利用前面学习的 HBase 的导入数据的知识将数据导入到 HBase 中，文件以“,”分割。表名为 goodinfo indexinfo。

38.3.4　建立关键词索引

从 38.3.3 节扒下来的数据只是存放在 HBase 中 goodinfo 表中，这个时候没有建立索引，我们在查询的时候是从索引表中根据关键词检索出跟关键词有关的 URL，并按照出现的次数按照相关性大小排序。所以我们需要将扒下来的数据建立索引，然后才能正确地使用。

在创建索引之前还需要做一些准备工作。因为创建索引的时候需要根据全文做分词提取关键词，然后根据关键词和链接的关系才能正确建立索引，所以需要分词工具的支持。分词工具支持的 jdk 为 jdk1.8。我们在创建索引之前需要把我们的 Hadoop 所依赖的 jdk 版本设置为 jdk1.8。修改步骤如下。

先解压 ~/data/38/jdk-8u121-linux-x64.tar.gz 文件。解压命令：

```
[root@client 38]# tar -zxvf  jdk-8u121-linux-x64.tar.gz
```

该命令将 jdk1.8 的压缩包直接解压在 38 目录下。

然后修改 Hadoop 集群中的配置文件 hadoop-env.sh（所有机器均需要修改）。修改命令如下：

```
[root@master hadoop]# vim /usr/cstor/hadoop/etc/hadoop/hadoop-env.sh
```

修改内容如下：

```
export JAVA_HOME=/root/data/38/jdk1.8.0_12
```

然后重启 Hadoop 集群即可。

我们建立索引的方式采用的是 MapReduce 的方法。我们将 Createindex 这个项目打包，然后向 Hadoop 中提交这个任务即可。Createindex 会根据商品的描述的分词结果建立关键词对 URL 的索引。所以我们还需要在建立分词的时候调用分词的数据，分词的数据存放在/root/data/38/data。你需要将这个文件夹复制到运行 jar 的机器的根目录下。

复制命令如下。

```
[root@client ~]# cp -r /root/data/38/data /
```

向 Hadoop 提交任务的命令如下：

```
[root@client hadoop]# bin/hadoop jar createindex.jar
```

38.3.5 关键词搜索

通过浏览器访问 TaobaoIndex.jsp，输入搜索的内容，并输入想选取的项数，然后点击检索，就会弹出你想了解的关键词的数据，如图 38-4 所示。

图 38-4　关键词搜索

38.4 实验结果

（1）扒取结果类似如图 38-5 所示。

图 38-5　扒取结果

（2）建立索引任务成功如图 38-6 所示。

```
16/12/14 11:28:21 INFO mapreduce.Job: Job job_1481094652573_0025 completed successfully
16/12/14 11:28:21 INFO mapreduce.Job: Counters: 49
        File System Counters
                FILE: Number of bytes read=880170
                FILE: Number of bytes written=2079445
                FILE: Number of read operations=0
                FILE: Number of large read operations=0
                FILE: Number of write operations=0
                HDFS: Number of bytes read=71
                HDFS: Number of bytes written=0
                HDFS: Number of read operations=1
                HDFS: Number of large read operations=0
                HDFS: Number of write operations=0
        Job Counters
                Launched map tasks=1
                Launched reduce tasks=1
                Data-local map tasks=1
                Total time spent by all maps in occupied slots (ms)=44770
                Total time spent by all reduces in occupied slots (ms)=23028
                Total time spent by all map tasks (ms)=44770
                Total time spent by all reduce tasks (ms)=11514
                Total vcore-seconds taken by all map tasks=44770
                Total vcore-seconds taken by all reduce tasks=11514
                Total megabyte-seconds taken by all map tasks=183377920
                Total megabyte-seconds taken by all reduce tasks=94322688
        Map-Reduce Framework
                Map input records=725
                Map output records=10948
                Map output bytes=858268
                Map output materialized bytes=880170
                Input split bytes=71
                Combine input records=0
                Combine output records=0
                Reduce input groups=1891
                Reduce shuffle bytes=880170
                Reduce input records=10948
                Reduce output records=1890
                Spilled Records=21896
                Shuffled Maps =1
                Failed Shuffles=0
                Merged Map outputs=1
                GC time elapsed (ms)=28505
                CPU time spent (ms)=217280
                Physical memory (bytes) snapshot=3860893696
                Virtual memory (bytes) snapshot=14571696128
                Total committed heap usage (bytes)=3653238784
        Shuffle Errors
                BAD_ID=0
                CONNECTION=0
                IO_ERROR=0
                WRONG_LENGTH=0
                WRONG_MAP=0
                WRONG_REDUCE=0
```

图 38-6　建立索引任务成功

（3）搜索结果如图 38-7 所示。

包邮　15　条　检索

共获取数据 15 条,搜索词为 包邮

#	商品名	价格	交易量	店铺	所在地
1	正品包邮 Homade 游戏机 碟盘 药丸 相机 个性手机壳 iphone4 4s	¥18.00	0人付款	telems	北京
2	躺椅配件大全 折叠躺椅子休闲椅ABS杯托 无味茶杯架 双开槽包邮	¥31.80	0人付款	changkonghao	浙江 丽水
3	特价包邮茶杯架用品椅子鱼箱大多功能桶钓箱钓鱼箱台钓箱渔具	¥148.00	0人付款	亿诚商贸会所	浙江 杭州
4	可折叠 不锈钢茶杯架 钓鱼椅杯架 茶杯座渔具 椅子配件包邮	¥26.00	0人付款	你我他841007	广东 广州
5	可折叠 不锈钢茶杯架 钓鱼椅杯架 茶杯座渔具 椅子配件包邮729	¥25.90	0人付款	一和商贸公司	浙江 金华
6	可折叠不锈钢茶杯架 钓鱼椅杯架 茶杯座垂钓渔具椅子配件包邮特价	¥19.00	0人付款	zhw108108	浙江 金华
7	可折叠 不锈钢茶杯架 钓鱼椅杯架 茶杯座渔具 椅子配件包邮729	¥56.90	0人付款	泉州巴藤服饰有限公司	浙江 金华
8	中国风故宫iphone7手机壳6s苹果7plus般若心经软壳浮雕防摔全包邮	¥45.00	64人付款	h5650377	山西 晋中
9	可折叠 不锈钢茶杯架 钓鱼椅杯架 茶杯座渔具 椅子配件包邮	¥190.00	0人付款	航航名店	浙江 金华
10	包邮希捷 2T 2TB 台式机 硬盘 监控 安防 专用 家用存储2000g串口	¥335.00	75人付款	weizhong6348269	广东 深圳
11	特价 包邮 7200转 1TB 台式机硬盘1000G硬盘 串口1T监控专用硬盘	¥245.00	151人付款	weizhong6348269	广东 深圳
12	包邮WD/西部数据 WD10JPLX 1TB黑盘1T 32M笔记本电脑硬盘7200转	¥482.00	39人付款	京美数码专营店	上海
13	可折叠 不锈钢茶杯架 钓鱼椅杯架 茶杯座渔具 椅子配件包邮	¥31.65	0人付款	张黄云9098	浙江 金华
14	可折叠 不锈钢茶杯架 钓鱼椅杯架 茶杯座渔具 椅子配件包邮	¥45.90	0人付款	maomao529529	上海
15	包邮希捷西数500G/1TB台式机硬盘机械硬盘二手硬盘监控硬盘7200转	¥160.00	215人付款	云轩电脑	广东 深圳

图 38-7　搜索结果展示

实验三十九　综合实战：推荐系统

39.1　实验目的

了解常用的基于矩阵分解的协同过滤推荐算法的基本原理，掌握 Spark MLlib 中对基于模型的协同过滤算法的封装函数的使用，对 Spark 中机器学习模块内容加深理解。

39.2　实验要求

1. 实验提供数据集，包括用户数据、电影数据、电影评分数据以及我的评分数据；
2. 根据提供的电影评分数据，利用 Spark 进行训练，得到一个最佳推荐模型；
3. 用实际数据和平均值这两方面评价该模型的准确度；
4. 根据我的评分数据向我推荐 10 部电影。

39.3　实验步骤

39.3.1　试验原理概述

协同过滤算法按照数据使用，可以分为：

（1）基于用户（UserCF）；

（2）基于商品（ItemCF）；

（3）基于模型（ModelCF）。

按照模型，又可以分为：

（1）最近邻模型：基于距离的协同过滤算法；

（2）Latent Factor Mode（SVD）：基于矩阵分解的模型；

（3）Graph：图模型，社会网络图模型。

本次实验，使用的协同过滤算法是基于矩阵分解的模型（其算法见图 39-1），就是基于样本的用户喜好信息，训练一个推荐模型，然后根据实时的用户喜好的信息进行预测，计算推荐。

交替最小二乘法（Alternating Least Squares，ALS），该方法常用于基于矩阵分解的推荐系统中。对于一个 ***R***（观众对电影的一个评价矩阵）可以分解为 ***U***（观众的特征矩阵）和 ***V***（电影的特征矩阵），在这个矩阵分解的过程中，评分缺失项得到了填充，也就是说我们可以基于这个填充的评分来给用户最适合的商品推荐。

MLlib 支持基于模型的协同过滤算法，其中 user 和 product 对应图中的 user 和 movie，user 和 product 之间有一些隐藏因子。MLlib 使用 ALS 来学习得到这些潜在因子。

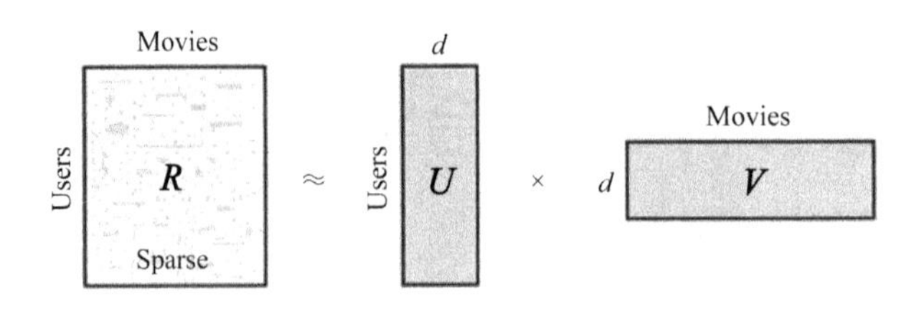

图 39-1　基于矩阵分解的模型算法图

图 39-1 中原始矩阵 ***R*** 可能是非常稀疏的，但乘积 ***UV*** 是稠密的，即使该矩阵存在非零元素，非零元素的数量也非常少。因此模型只是对 ***R*** 的一种近似。原始矩阵 ***R*** 中大量元素是缺失的（元素值为 0），算法为这些缺失元素生成（补全）了一个值，从这个角度来讲，我们可以把算法称为模型。根据这个补全的矩阵，我们就可以知道 user 也就知道了 movies，或者知道 movie 也就知道了 users，这就是以下实验推荐算法的基本原理。

39.3.2　数据集准备

构建模型的第一步是了解数据，对数据进行解析或转换，以便在 Spark 中作分析。Spark MLlib 的 ALS 算法要求用户和产品的 ID 必须都是数值型，并且是 32 位非负整数，以下准备的数据集完全符合 Spark MLlib 的 ALS 算法要求，不必进行转换，可直接使用。

在本地目录/root/data/39/movie 下有本次实验数据集，文件列表如图 39-2 所示。

```
-rw-r--r-- 1 root root   171308 Dec  9 13:44 movies.dat
-rw-r--r-- 1 root root 24594131 Dec  9 13:44 ratings.dat
-rw-r--r-- 1 root root     5197 Dec  9 13:44 README
-rw-r--r-- 1 root root      251 Dec  9 13:44 test.dat
-rw-r--r-- 1 root root   134368 Dec  9 13:44 users.dat
```

图 39-2　文件列表

各文件数据格式如下（详细见 README 文件）：

（1）用户数据（users.dat）

用户 ID::性别::年龄::职业编号::邮编。

```
6031::F::18::0::45123
6032::M::45::7::55108
6033::M::50::13::78232
6034::M::25::14::94117
6035::F::25::1::78734
6036::F::25::15::32603
6037::F::45::1::76006
6038::F::56::1::14706
6039::F::45::0::01060
6040::M::25::6::11106
```

（2）电影数据（movies.dat）

电影 ID::电影名称::电影种类。

```
3943::Bamboozled (2000)::Comedy
3944::Bootmen (2000)::Comedy|Drama
3945::Digimon: The Movie (2000)::Adventure|Animation|Children′s
3946::Get Carter (2000)::Action|Drama|Thriller
3947::Get Carter (1971)::Thriller
3948::Meet the Parents (2000)::Comedy
3949::Requiem for a Dream (2000)::Drama
3950::Tigerland (2000)::Drama
3951::Two Family House (2000)::Drama
3952::Contender, The (2000)::Drama|Thriller
```

（3）评分数据（ratings.dat）

用户 ID::电影 ID::评分::时间。

```
6040::2022::5::956716207
6040::2028::5::956704519
6040::1080::4::957717322
6040::1089::4::956704996
6040::1090::3::956715518
6040::1091::1::956716541
6040::1094::5::956704887
6040::562::5::956704746
6040::1096::4::956715648
6040::1097::4::956715569
```

（4）我的评分数据（test.dat）

用户 ID::电影 ID::评分::时间。

```
0::780::4::1409495135
0::590::3::1409495135
0::1210::4::1409495135
0::648::5::1409495135
0::344::3::1409495135
0::165::4::1409495135
0::153::5::1409495135
0::597::4::1409495135
0::1580::5::1409495135
```

将以上数据文件上传到 HDFS 文件系统：

```
cd /usr/cstor/hadoop/bin
hdfs dfs -copyFromLocal /root/data/39/movie/ /
```

39.3.3　代码实现

为防止 Shell 端 INFO 日志刷屏，影响查看打印信息，修改打印日志级别，进入 Spark 安装的 conf 目录下，将 log4j.properties.template 文件复制一份，命名为

log4j.properties 文件，然后将文件如下配置项：

```
log4j.rootCategory=WARN, console
```

进入 Spark 安装目录下 bin 目录，启动 spark-shell：

```
cd /usr/cstor/spark/bin
./spark-shell --master spark://master:7077
```

具体代码如下：

```
/** 导入 Spark 机器学习推荐算法相关包 **/
import org.apache.spark.mllib.recommendation.{ALS, Rating, MatrixFactorizationModel}
import org.apache.spark.rdd.RDD

/** 定义函数，校验集预测数据和实际数据之间的均方根误差，后面会调用此函数 **/
def computeRmse(model:MatrixFactorizationModel,data:RDD[Rating],n:Long):Double = {
  val predictions:RDD[Rating] = model.predict((data.map(x => (x.user,x.product))))
  val predictionsAndRatings = predictions.map{ x =>((x.user,x.product),x.rating)}
    .join(data.map(x => ((x.user,x.product),x.rating))).values
  math.sqrt(predictionsAndRatings.map( x => (x._1 - x._2) * (x._1 - x._2)).reduce(_+_)/n)
}

/** 加载数据 **/
//1. 我的评分数据(test.dat),转成 Rating 格式，即用户 id，电影 id，评分
  val myRatingsRDD = sc.textFile("/movie/test.dat").map {
    line =>
      val fields = line.split("::")
      // format: Rating(userId, movieId, rating)
      Rating(fields(0).toInt, fields(1).toInt, fields(2).toDouble)
  }
//2. 样本评分数据(ratings.dat)，其中最后一列 Timestamp 取除 10 的余数作为 key，Rating 为值，
 即(Int，Rating)，以备后续数据切分
  val ratings = sc.textFile("/movie/ratings.dat").map {
    line =>
      val fields = line.split("::")
      // format: (timestamp % 10, Rating(userId, movieId, rating))
      (fields(3).toLong % 10, Rating(fields(0).toInt, fields(1).toInt, fields(2).toDouble))
  }
  //3. 电影数据(movies.dat)(电影 ID->电影标题)
  val movies = sc.textFile("/movie/movies.dat").map {
    line =>
      val fields = line.split("::")
      // format: (movieId, movieName)
      (fields(0).toInt, fields(1))
  }.collect().toMap

  /** 统计所有用户数量和电影数量以及用户对电影的评分数目 **/
```

```
        val numRatings = ratings.count()
        val numUsers = ratings.map(_._2.user).distinct().count()
        val numMovies = ratings.map(_._2.product).distinct().count()
        println("total number of rating data: " + numRatings)
        println("number of users participating in the score: " + numUsers)
        println("number of participating movie data: " + numMovies)

    /** 将样本评分表以 key 值切分成 3 个部分，分别用于训练（60%，并加入我的评分数据）、
     校验（20%）以及测试（20%） **/
        //定义分区数，即数据并行度
         val numPartitions = 4
        //因为以下数据在计算过程中要多次应用到，所以 cache 到内存
        //训练数据集，包含我的评分数据
        val training = ratings.filter(x => x._1 < 6).values.union(myRatingsRDD). repartition(numPartitions).
persist()
        //验证数据集
        val validation = ratings.filter(x => x._1 >= 6 && x._1 < 8).values.repartition(numPartitions).persist()
        //测试数据集
        val test = ratings.filter(x => x._1 >= 8).values.persist()
        //统计各数据集数量
        val numTraining = training.count()
        val numValidation = validation.count()
        val numTest = test.count()
        println("the number of scoring data for training) (including my score data): " + numTraining)
        println("number of rating data as validation: " + numValidation)
        println("number of rating data as a test: " + numTest)

    /** 训练不同参数下的模型，获取最佳模型 **/
    //设置训练参数及最佳模型初始化值
    //模型的潜在因素的个数，即 U 和 V 矩阵的列数，也叫矩阵的阶
    val ranks = List(8, 12)
    //标准的过拟合参数
    val lambdas = List(0.1, 10.0)
    //矩阵分解迭代次数，次数越多花费时间越长，分解的结果也可能会更好
        val numIters = List(10, 20)
        var bestModel: Option[MatrixFactorizationModel] = None
        var bestValidationRmse = Double.MaxValue
        var bestRank = 0
        var bestLambda = -1.0
        var bestNumIter = -1
        //根据设定的训练参数对训练数据集进行训练
    for (rank <- ranks; lambda <- lambdas; numIter <- numIters) {
      //计算模型
          val model = ALS.train(training, rank, numIter, lambda)
```

```
        //计算针对校验集的预测数据和实际数据之间的均方根误差
        val validationRmse = computeRmse(model, validation, numValidation)
        println("Root mean square: " + validationRmse + " Parameter: --rank = "
          + rank + " --lambda = " + lambda + " --numIter = " + numIter + ".")

        //均方根误差最小的为最佳模型
        if (validationRmse < bestValidationRmse) {
          bestModel = Some(model)
          bestValidationRmse = validationRmse
          bestRank = rank
          bestLambda = lambda
          bestNumIter = numIter
        }
      }

      /** 用训练的最佳模型预测评分并评估模型准确度 **/
      //训练完成后，用最佳模型预测测试集的评分，并计算和实际评分之间的均方根误差
（RMSE）
      val testRmse = computeRmse(bestModel.get, test, numTest)
      println("Optimal model parameters --rank = " + bestRank + " --lambda = " + bestLambda
                          + " --numIter = " + bestNumIter +
            " \nThe root mean square between the predicted data and the real data under the optimal
model: " + testRmse + ".")

   //创建一个用均值预测的评分，并与最好的模型进行比较，
    这个 mean（）方法在 DoubleRDDFunctions 中，求平均值
      val meanRating = training.union(validation).map(_.rating).mean
   val baselineRmse = math.sqrt(test.map(x => (meanRating - x.rating) * (meanRating - x.rating))
   .reduce(_ + _) / numTest)
      println("Root mean square between mean prediction data and real data: " + baselineRmse + ".")
      val improvement = (baselineRmse - testRmse) / baselineRmse * 100
   println("The accuracy of the prediction data of the best model with respect to the mean prediction data: " +
"%1.2f".format(improvement) + "%.")

      //向我推荐十部最感兴趣的电影
      val recommendations = bestModel.get.recommendProducts(0,10)
      //打印推荐结果
      var i = 1
      println("10 films recommended to me:")
      recommendations.foreach { r =>
        println("%2d".format(i) + ": " + movies(r.product))
        i += 1
      }
```

39.4 实验结果

代码执行过程中打印日志信息，如图 39-3～图 39-9 所示。

```
scala>     println("total number of rating data: " + numRatings)
total number of rating data: 1000209

scala>     println("number of users participating in the score: " + numUsers)
number of users participating in the score: 6040

scala>     println("number of participating movie data: " + numMovies)
number of participating movie data: 3706
```

图 39-3　所有数据数量统计

```
scala>     println("the number of scoring data for training) (including my score data): " + numTraining)
the number of scoring data for training) (including my score data): 602252

scala>     println("number of rating data as validation: " + numValidation)
number of rating data as validation: 198919

scala>     println("number of rating data as a test: " + numTest)
number of rating data as a test: 199049
```

图 39-4　评分数据切分的各数据集统计

```
Root mean square: 0.8781910506804202 Parameter: --rank = 8 --lambda = 0.1 --numIter = 10.
Root mean square: 0.8728043765574968 Parameter: --rank = 8 --lambda = 0.1 --numIter = 20.
Root mean square: 3.7558695311242833 Parameter: --rank = 8 --lambda = 10.0 --numIter = 10.
Root mean square: 3.7558695311242833 Parameter: --rank = 8 --lambda = 10.0 --numIter = 20.
Root mean square: 0.8762686398271575 Parameter: --rank = 12 --lambda = 0.1 --numIter = 10.
Root mean square: 0.870803465191469 Parameter: --rank = 12 --lambda = 0.1 --numIter = 20.
Root mean square: 3.7558695311242833 Parameter: --rank = 12 --lambda = 10.0 --numIter = 10.
Root mean square: 3.7558695311242833 Parameter: --rank = 12 --lambda = 10.0 --numIter = 20.
```

图 39-5　训练时的参数及对应的误差

```
Optimal model parameters --rank = 12 --lambda = 0.1 --numIter = 20
The root mean square between the predicted data and the real data under the optimal model: 0.8687969796603212.
```

图 39-6　最佳模型的参数及对应的误差

```
scala> println("Root mean square between mean prediction data and real data: " + baselineRmse + ".")
Root mean square between mean prediction data and real data: 1.1135136649013289.
```

图 39-7　均值预测的误差

```
scala> println("The accuracy of the prediction data of the best model with respect to the mean prediction data: " + "%1.2f".format(improvement) + "%.")
The accuracy of the prediction data of the best model with respect to the mean prediction data: 21.98%.
```

图 39-8　最佳模型预测和均值预测的比较

```
scala> println("10 films recommended to me:")
10 films recommended to me:

scala>     recommendations.foreach { r =>
     |       println("%2d".format(i) + ": " + movies(r.product))
     |       i += 1
     |     }
 1: Love Serenade (1996)
 2: Chushingura (1962)
 3: Some Mother's Son (1996)
 4: Fear of a Black Hat (1993)
 5: Raiders of the Lost Ark (1981)
 6: Very Thought of You, The (1998)
 7: Ayn Rand: A Sense of Life (1997)
 8: Braindead (1992)
 9: First Love, Last Rites (1997)
10: Die Hard (1988)
```

图 39-9　最佳模型下向我推荐的电影

实验四十 综合实战：环境大数据

40.1 实验目的

1. 分析环境数据文件；
2. 编写解析环境数据文件并进行统计的代码；
3. 进行递归 MapReduce。

40.2 实验要求

要求实验结束时，每位学生均已在 client 服务器上运行，从北京 2016 年 1～6 月这半年间的历史天气和空气质量数据文件中分析出的环境统计结果，包括月平均气温、空气质量分布情况等。

40.3 实验原理

近年来，由于雾霾问题的持续发酵，越来越多的人开始关注城市相关的环境数据，包括空气质量数据、天气数据等。

如果每小时记录一次城市的天气实况和空气质量实况信息，则每个城市每天都会产生 24 条环境数据，全国所有 2500 多个城市如果均如此进行记录，那每天产生的数据量将达到 6 万多条，每年则会产生 2190 万条记录，已经可以称得上是环境大数据。

对于这些原始监测数据，我们可以根据时间的维度来进行统计，从而得出与该城市相关的日度及月度平均气温、空气质量优良及污染天数等，从而为研究空气污染物扩散条件提供有力的数据支持。

本实验中选取了北京 2016 年 1～6 月，这半年间每小时天气和空气质量数据（未取到数据的字段填充“N/A”），利用 MapReduce 来统计月度平均气温和半年内空气质量为优、良、轻度污染、中度污染、重度污染和严重污染的天数。

40.4 实验步骤

40.4.1 分析数据文件

在 client 服务器上执行下列命令，查看环境数据文件 beijing.txt（见图 40-1），路径在/root/data/40 目录下。

```
[root@client ~]# more /root/data/40/beijing.txt
```

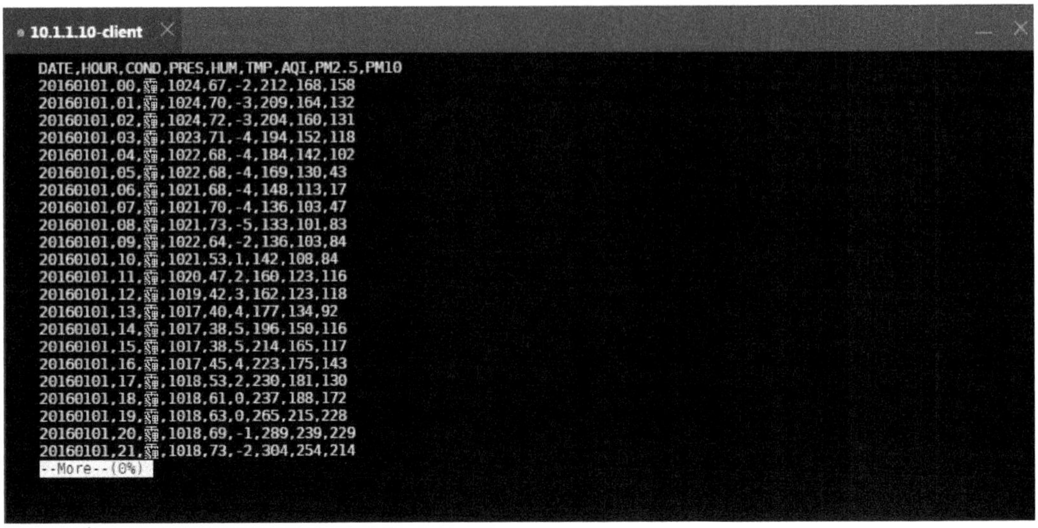

```
10.1.1.10-client
DATE,HOUR,COND,PRES,HUM,TMP,AQI,PM2.5,PM10
20160101,00,霾,1024,67,-2,212,168,158
20160101,01,霾,1024,70,-3,209,164,132
20160101,02,霾,1024,72,-3,204,160,131
20160101,03,霾,1023,71,-4,194,152,118
20160101,04,霾,1022,68,-4,184,142,102
20160101,05,霾,1022,68,-4,169,130,43
20160101,06,霾,1021,68,-4,148,113,17
20160101,07,霾,1021,70,-4,136,103,47
20160101,08,霾,1021,73,-5,133,101,83
20160101,09,霾,1022,64,-2,136,103,84
20160101,10,霾,1021,53,1,142,108,84
20160101,11,霾,1020,47,2,160,123,116
20160101,12,霾,1019,42,3,162,123,118
20160101,13,霾,1017,40,4,177,134,92
20160101,14,霾,1017,38,5,196,150,116
20160101,15,霾,1017,38,5,214,165,117
20160101,16,霾,1017,45,4,223,175,143
20160101,17,霾,1018,53,2,230,181,130
20160101,18,霾,1018,61,0,237,188,172
20160101,19,霾,1018,63,0,265,215,228
20160101,20,霾,1018,69,-1,289,239,229
20160101,21,霾,1018,73,-2,304,254,214
--More--(0%)
```

图 40-1　环境数据文件格式

从图 40-1 种可以看到，我们需要关心的数据有第一列 DATE、第二列 HOUR、第六列 TMP 和第七列 AQI 的数据。

40.4.2　将数据文件上传至 HDFS

在 client 上传 beijing.txt 到 HDFS 的/input 目录上。

```
[root@client ~]# hadoop fs -mkdir /input
[root@client ~]# hadoop fs -put ~/data/40/beijing.txt /input
```

40.4.3　编写月平均气温统计程序

在 Eclipse 上新建 MapReduce 项目，命名为 TmpStat，在 src 目录下新建文件 TmpStat.java，并键入如下代码。

```
import java.io.IOException;

import org.apache.hadoop.conf.Configuration;
import org.apache.hadoop.fs.Path;
import org.apache.hadoop.io.IntWritable;
import org.apache.hadoop.io.Text;
import org.apache.hadoop.mapreduce.Job;
import org.apache.hadoop.mapreduce.Mapper;
import org.apache.hadoop.mapreduce.Reducer;
import org.apache.hadoop.mapreduce.lib.input.TextInputFormat;
import org.apache.hadoop.mapreduce.lib.output.TextOutputFormat;
import org.apache.hadoop.mapreduce.lib.partition.HashPartitioner;
```

```
public class TmpStat
{
    public static class StatMapper extends Mapper<Object, Text, Text, IntWritable>
    {
        private IntWritable intValue = new IntWritable();
        private Text dateKey = new Text();

        public void map(Object key, Text value, Context context)
                throws IOException, InterruptedException
        {
            String[] items = value.toString().split(",");

            String date = items[0];
            String tmp = items[5];

            if(!"DATE".equals(date) && !"N/A".equals(tmp))
            {//排除第一行说明以及未取到数据的行
                dateKey.set(date.substring(0, 6));
                intValue.set(Integer.parseInt(tmp));
                context.write(dateKey, intValue);
            }
        }
    }

    public static class StatReducer extends Reducer<Text, IntWritable, Text, IntWritable>
    {
        private IntWritable result = new IntWritable();
        public void reduce(Text key, Iterable<IntWritable> values, Context context)
                throws IOException, InterruptedException
        {
            int tmp_sum = 0;
            int count = 0;

            for(IntWritable val : values)
            {
                tmp_sum += val.get();
                count++;
            }

            int tmp_avg = tmp_sum/count;
            result.set(tmp_avg);
            context.write(key, result);
        }
```

```
    }

    public static void main(String args[])
            throws IOException, ClassNotFoundException, InterruptedException
    {
        Configuration conf = new Configuration();
        Job job = new Job(conf, "MonthlyAvgTmpStat");
        job.setInputFormatClass(TextInputFormat.class);
        TextInputFormat.setInputPaths(job, args[0]);
        job.setJarByClass(TmpStat.class);
        job.setMapperClass(StatMapper.class);
        job.setMapOutputKeyClass(Text.class);
        job.setMapOutputValueClass(IntWritable.class);
        job.setPartitionerClass(HashPartitioner.class);
        job.setReducerClass(StatReducer.class);
        job.setNumReduceTasks(Integer.parseInt(args[2]));
        job.setOutputKeyClass(Text.class);
        job.setOutputValueClass(IntWritable.class);
        job.setOutputFormatClass(TextOutputFormat.class);
        TextOutputFormat.setOutputPath(job, new Path(args[1]));
        System.exit(job.waitForCompletion(true) ? 0 : 1);
    }
}
```

使用 Eclipse 软件将 TmpStat 项目导出成 jar 文件，指定主类为 TmpStat，命名为 tmpstat.jar，并上传至 client 服务器上。

40.4.4 查看月平均气温统计结果

在 client 上执行 tmpstat.jar，指定输出目录为/monthlyavgtmp，reducer 数量为 1，如图 40-2 所示。

```
[root@client ~]# hadoop jar tmpstat.jar /input /monthlyavgtmp 1
```

图 40-2 运行 tmpstat.jar

在 client 上查看统计结果，如图 40-3 所示。

```
[root@client ~]# hadoop fs -ls /monthlyavgtmp
[root@client ~]# hadoop fs -cat /monthlyavgtmp/part-r-00000
```

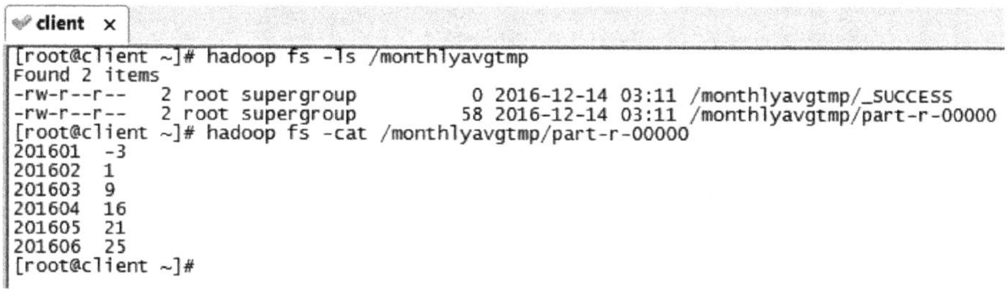

```
client ×
[root@client ~]# hadoop fs -ls /monthlyavgtmp
Found 2 items
-rw-r--r--   2 root supergroup          0 2016-12-14 03:11 /monthlyavgtmp/_SUCCESS
-rw-r--r--   2 root supergroup         58 2016-12-14 03:11 /monthlyavgtmp/part-r-00000
[root@client ~]# hadoop fs -cat /monthlyavgtmp/part-r-00000
201601  -3
201602  1
201603  9
201604  16
201605  21
201606  25
[root@client ~]#
```

图 40-3　查看月平均气温统计结果

40.4.5　编写每日空气质量统计程序

在 Eclipse 上新建 MapReduce 项目，命名为 AqiStatDaily，在 src 目录下新建文件 AqiStatDaily.java，并键入如下代码。

```
import java.io.IOException;

import org.apache.hadoop.conf.Configuration;
import org.apache.hadoop.fs.Path;
import org.apache.hadoop.io.IntWritable;
import org.apache.hadoop.io.Text;
import org.apache.hadoop.mapreduce.Job;
import org.apache.hadoop.mapreduce.Mapper;
import org.apache.hadoop.mapreduce.Reducer;
import org.apache.hadoop.mapreduce.lib.input.TextInputFormat;
import org.apache.hadoop.mapreduce.lib.output.TextOutputFormat;
import org.apache.hadoop.mapreduce.lib.partition.HashPartitioner;

public class AqiStatDaily
{
  public static class StatMapper extends Mapper<Object, Text, Text, IntWritable>
  {
    private IntWritable intValue = new IntWritable();
    private Text dateKey = new Text();

    public void map(Object key, Text value, Context context)
          throws IOException, InterruptedException
    {
      String[] items = value.toString().split(",");

      String date = items[0];
```

```
            String aqi = items[6];

            if(!"DATE".equals(date) && !"N/A".equals(aqi))
            {
                dateKey.set(date);
                intValue.set(Integer.parseInt(aqi));
                context.write(dateKey, intValue);
            }
        }
    }

    public static class StatReducer extends Reducer<Text, IntWritable, Text, IntWritable>
    {
        private IntWritable result = new IntWritable();
        public void reduce(Text key, Iterable<IntWritable> values, Context context)
                throws IOException, InterruptedException
        {
            int aqi_sum = 0;
            int count = 0;

            for(IntWritable val : values)
            {
                aqi_sum += val.get();
                count++;
            }

            int aqi_avg = aqi_sum/count;
            result.set(aqi_avg);
            context.write(key, result);
        }
    }

    public static void main(String args[])
            throws IOException, ClassNotFoundException, InterruptedException
    {
        Configuration conf = new Configuration();
        Job job = new Job(conf, "AqiStatDaily");
        job.setInputFormatClass(TextInputFormat.class);
        TextInputFormat.setInputPaths(job, args[0]);
        job.setJarByClass(AqiStatDaily.class);
        job.setMapperClass(StatMapper.class);
        job.setMapOutputKeyClass(Text.class);
        job.setMapOutputValueClass(IntWritable.class);
        job.setPartitionerClass(HashPartitioner.class);
```

```
        job.setReducerClass(StatReducer.class);
        job.setNumReduceTasks(Integer.parseInt(args[2]));
        job.setOutputKeyClass(Text.class);
        job.setOutputValueClass(IntWritable.class);
        job.setOutputFormatClass(TextOutputFormat.class);
        TextOutputFormat.setOutputPath(job, new Path(args[1]));
        System.exit(job.waitForCompletion(true) ? 0 : 1);
    }
}
```

使用 Eclipse 软件将 AqiStatDaily 项目导出成 jar 文件，指定主类为 AqiStatDaily，命名为 aqistatdaily.jar，并上传至 client 服务器上。

40.4.6　查看每日空气质量统计结果

在 client 上执行 aqistatdaily.jar，指定输出目录为/aqidaily，reducer 数量为 3。如图 40-4 所示。

```
[root@client ~]# hadoop jar aqistatdaily.jar /input /aqidaily 3
```

```
client ×
16/12/14 03:44:20 INFO mapred.LocalJobRunner: 1 / 1 copied.
16/12/14 03:44:20 INFO mapred.Task: Task:attempt_local983719235_0001_r_000002_0 is done. And is in the process of committing
16/12/14 03:44:20 INFO mapred.LocalJobRunner: 1 / 1 copied.
16/12/14 03:44:20 INFO mapred.Task: Task attempt_local983719235_0001_r_000002_0 is allowed to commit now
16/12/14 03:44:20 INFO output.FileOutputCommitter: Saved output of task 'attempt_local983719235_0001_r_000002_0' to hdfs://master:8020/aqidaily/_temporary/0/task_local983719235_0001_r_000002
16/12/14 03:44:20 INFO mapred.LocalJobRunner: reduce > reduce
16/12/14 03:44:20 INFO mapred.Task: Task 'attempt_local983719235_0001_r_000002_0' done.
16/12/14 03:44:20 INFO mapred.LocalJobRunner: Finishing task: attempt_local983719235_0001_r_000002_0
16/12/14 03:44:20 INFO mapred.LocalJobRunner: reduce task executor complete.
16/12/14 03:44:20 INFO mapreduce.Job:  map 100% reduce 100%
16/12/14 03:44:20 INFO mapreduce.Job: Job job_local983719235_0001 completed successfully
16/12/14 03:44:20 INFO mapreduce.Job: Counters: 35
        File System Counters
                FILE: Number of bytes read=250508
                FILE: Number of bytes written=1441648
                FILE: Number of read operations=0
                FILE: Number of large read operations=0
                FILE: Number of write operations=0
                HDFS: Number of bytes read=683852
                HDFS: Number of bytes written=4510
                HDFS: Number of read operations=38
                HDFS: Number of large read operations=0
                HDFS: Number of write operations=18
        Map-Reduce Framework
                Map input records=4369
                Map output records=3438
                Map output bytes=44694
                Map output materialized bytes=51588
                Input split bytes=101
                Combine input records=0
                Combine output records=0
                Reduce input groups=182
                Reduce shuffle bytes=51588
                Reduce input records=3438
                Reduce output records=182
                Spilled Records=6876
                Shuffled Maps =3
                Failed Shuffles=0
                Merged Map outputs=3
                GC time elapsed (ms)=19
                Total committed heap usage (bytes)=2059403264
        Shuffle Errors
                BAD_ID=0
                CONNECTION=0
                IO_ERROR=0
                WRONG_LENGTH=0
                WRONG_MAP=0
                WRONG_REDUCE=0
        File Input Format Counters
                Bytes Read=170963
        File Output Format Counters
                Bytes Written=2250
[root@client ~]#
```

图 40-4　运行 aqistatdaily.jar

在 client 上查看统计结果文件，如图 40-5 所示。

```
[root@client ~]# hadoop fs -ls /aqidaily
```

```
client ×
[root@client ~]# hadoop fs -ls /aqidaily
Found 4 items
-rw-r--r--   2 root supergroup          0 2016-12-14 03:44 /aqidaily/_SUCCESS
-rw-r--r--   2 root supergroup        752 2016-12-14 03:44 /aqidaily/part-r-00000
-rw-r--r--   2 root supergroup        756 2016-12-14 03:44 /aqidaily/part-r-00001
-rw-r--r--   2 root supergroup        742 2016-12-14 03:44 /aqidaily/part-r-00002
[root@client ~]#
```

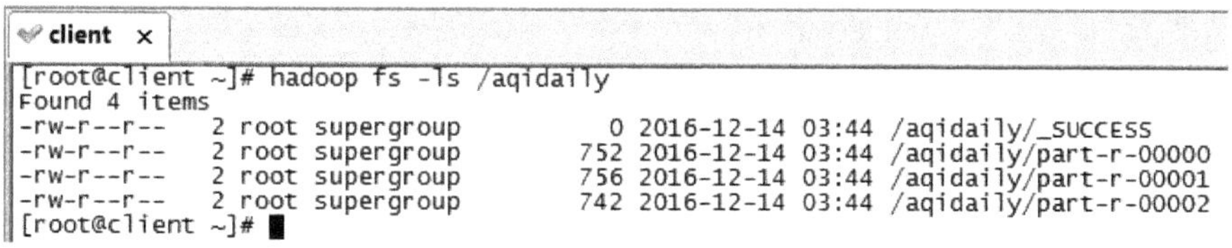

图 40-5　查看 aqistatdaily.jar 运行结果文件

可以看到，结果文件被分成了 3 个部分，依次查看这 3 个文件的内容，即可看到每天的空气质量统计结果数据，如图 40-6 所示。

```
[root@client ~]# hadoop fs -cat /aqidaily/p*
```

```
client ×
20160317        335
20160320        85
20160323        40
20160326        34
20160329        56
20160401        111
20160404        96
20160407        142
20160410        124
20160413        162
20160416        70
20160419        95
20160422        80
20160425        126
20160428        116
20160503        55
20160506        137
20160509        56
20160512        69
20160515        55
20160518        89
20160521        105
20160524        43
20160527        85
20160530        92
20160602        114
20160605        105
20160608        154
20160611        31
20160614        52
20160617        48
20160620        118
20160623        109
20160626        72
20160629        145
[root@client ~]#
```

图 40-6　查看每日空气质量统计结果

40.4.7　将每日空气质量统计文件进行整合

在 client 服务器上将每日空气质量统计结果保存到 aqidaily.txt。

```
[root@client ~]# hadoop fs -cat /aqidaily/part-r-00000 > aqidaily.txt
[root@client ~]# hadoop fs -cat /aqidaily/part-r-00001 >> aqidaily.txt
[root@client ~]# hadoop fs -cat /aqidaily/part-r-00002 >> aqidaily.txt
[root@client ~]# cat aqidaily.txt |wc -l
182
[root@client ~]#
```

在 HDFS 上创建/aqiinput 目录，并将 aqidaily.txt 上传至该目录下。

```
[root@client ~]# hadoop fs -mkdir /aqiinput
[root@client ~]# hadoop fs -put aqidaily.txt /aqiinput
```

40.4.8　编写各空气质量天数统计程序

在 Eclipse 上新建 MapReduce 项目，命名为 AqiStat，在 src 目录下新建文件 AqiStat.java，并键入如下代码。

```
import java.io.IOException;


import org.apache.hadoop.conf.Configuration;
```

```
import org.apache.hadoop.fs.Path;
import org.apache.hadoop.io.IntWritable;
import org.apache.hadoop.io.Text;
import org.apache.hadoop.mapreduce.Job;
import org.apache.hadoop.mapreduce.Mapper;
import org.apache.hadoop.mapreduce.Reducer;
import org.apache.hadoop.mapreduce.lib.input.TextInputFormat;
import org.apache.hadoop.mapreduce.lib.output.TextOutputFormat;
import org.apache.hadoop.mapreduce.lib.partition.HashPartitioner;

public class AqiStat
{
  public static final String GOOD = "优";
  public static final String MODERATE = "良";
  public static final String LIGHTLY_POLLUTED = "轻度污染";
  public static final String MODERATELY_POLLUTED = "中度污染";
  public static final String HEAVILY_POLLUTED = "重度污染";
  public static final String SEVERELY_POLLUTED = "严重污染";

  public static class StatMapper extends Mapper<Object, Text, Text, IntWritable>
  {
    private final static IntWritable one = new IntWritable(1);
    private Text cond = new Text();
    // map 方法，根据 AQI 值，将对应空气质量的天数加 1
    public void map(Object key, Text value, Context context)
            throws IOException, InterruptedException
    {
      String[] items = value.toString().split("\t");
      int aqi = Integer.parseInt(items[1]);

      if(aqi <= 50)
      {
        // 优
        cond.set(GOOD);
      }
      else if(aqi <= 100)
      {
        // 良
        cond.set(MODERATE);
      }
      else if(aqi <= 150)
      {
        // 轻度污染
        cond.set(LIGHTLY_POLLUTED);
```

```
        }
        else if(aqi <= 200)
        {
            // 中度污染
            cond.set(MODERATELY_POLLUTED);
        }
        else if(aqi <= 300)
        {
            // 重度污染
            cond.set(HEAVILY_POLLUTED);
        }
        else
        {
            // 严重污染
            cond.set(SEVERELY_POLLUTED);
        }

        context.write(cond, one);
    }
}
// 定义 reduce 类，对相同的空气质量状况，把它们<K,VList>中 VList 值全部相加
public static class StatReducer extends Reducer<Text, IntWritable, Text, IntWritable>
{
    private IntWritable result = new IntWritable();
    public void reduce(Text key, Iterable<IntWritable> values,Context context)
            throws IOException, InterruptedException
    {
        int sum = 0;
        for (IntWritable val : values)
        {
            sum += val.get();
        }
        result.set(sum);
        context.write(key, result);
    }
}

public static void main(String args[])
        throws IOException, ClassNotFoundException, InterruptedException
{
    Configuration conf = new Configuration();
    Job job = new Job(conf, "AqiStat");
    job.setInputFormatClass(TextInputFormat.class);
    TextInputFormat.setInputPaths(job, args[0]);
```

```
        job.setJarByClass(AqiStat.class);
        job.setMapperClass(StatMapper.class);
        job.setCombinerClass(StatReducer.class);
        job.setMapOutputKeyClass(Text.class);
        job.setMapOutputValueClass(IntWritable.class);
        job.setPartitionerClass(HashPartitioner.class);
        job.setReducerClass(StatReducer.class);
        job.setNumReduceTasks(Integer.parseInt(args[2]));
        job.setOutputKeyClass(Text.class);
        job.setOutputValueClass(IntWritable.class);
        job.setOutputFormatClass(TextOutputFormat.class);
        TextOutputFormat.setOutputPath(job, new Path(args[1]));
        System.exit(job.waitForCompletion(true) ? 0 : 1);
    }
}
```

使用 Eclipse 软件将 AqiStat 项目导出成 jar 文件，指定主类为 AqiStat，命名为 aqistat.jar，并上传至 client 服务器上。

40.4.9　查看各空气质量天数统计结果

在 client 上执行 aqistat.jar，指定输出目录为/aqioutput，reducer 数量为 1。如图 40-7 所示。

```
[root@client ~]# hadoop jar aqistat.jar /aqiinput /aqioutput 1
```

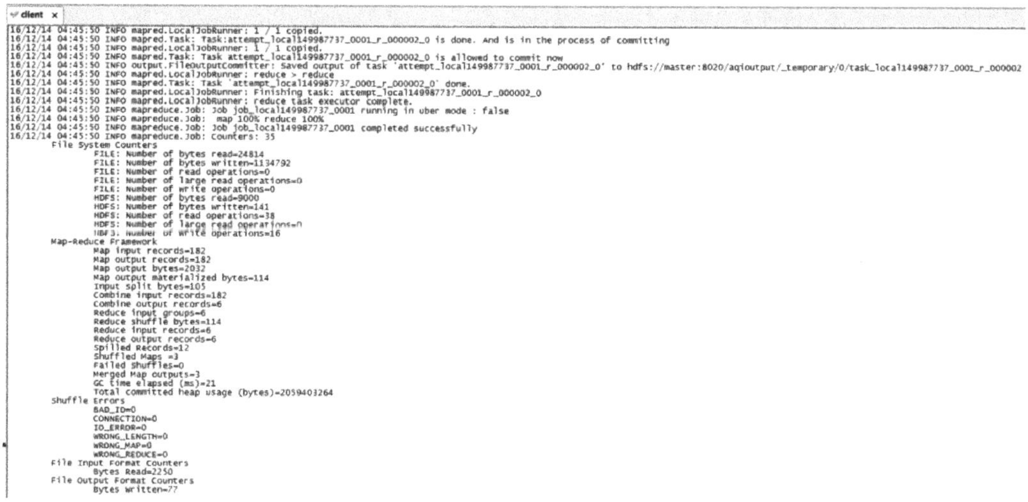

```
client ×
16/12/14 04:45:50 INFO mapred.LocalJobRunner: 1 / 1 copied.
16/12/14 04:45:50 INFO mapred.Task: Task:attempt_local149987737_0001_r_000002_0 is done. And is in the process of committing
16/12/14 04:45:50 INFO mapred.LocalJobRunner: 1 / 1 copied.
16/12/14 04:45:50 INFO mapred.Task: Task attempt_local149987737_0001_r_000002_0 is allowed to commit now
16/12/14 04:45:50 INFO output.FileOutputCommitter: Saved output of task 'attempt_local149987737_0001_r_000002_0' to hdfs://master:8020/aqioutput/_temporary/0/task_local149987737_0001_r_000002
16/12/14 04:45:50 INFO mapred.LocalJobRunner: reduce > reduce
16/12/14 04:45:50 INFO mapred.Task: Task 'attempt_local149987737_0001_r_000002_0' done.
16/12/14 04:45:50 INFO mapred.LocalJobRunner: Finishing task: attempt_local149987737_0001_r_000002_0
16/12/14 04:45:50 INFO mapred.LocalJobRunner: reduce task executor complete.
16/12/14 04:45:50 INFO mapreduce.Job: Job job_local149987737_0001 running in uber mode : false
16/12/14 04:45:50 INFO mapreduce.Job:  map 100% reduce 100%
16/12/14 04:45:50 INFO mapreduce.Job: Job job_local149987737_0001 completed successfully
16/12/14 04:45:50 INFO mapreduce.Job: Counters: 35
        File System Counters
                FILE: Number of bytes read=24814
                FILE: Number of bytes written=1134792
                FILE: Number of read operations=0
                FILE: Number of large read operations=0
                FILE: Number of write operations=0
                HDFS: Number of bytes read=9000
                HDFS: Number of bytes written=141
                HDFS: Number of read operations=38
                HDFS: Number of large read operations=0
                HDFS: Number of write operations=16
        Map-Reduce Framework
                Map input records=182
                Map output records=182
                Map output bytes=2032
                Map output materialized bytes=114
                Input split bytes=105
                Combine input records=182
                Combine output records=6
                Reduce input groups=6
                Reduce shuffle bytes=114
                Reduce input records=6
                Reduce output records=6
                Spilled Records=12
                Shuffled Maps =3
                Failed Shuffles=0
                Merged Map outputs=3
                GC time elapsed (ms)=21
                Total committed heap usage (bytes)=2059403264
        Shuffle Errors
                BAD_ID=0
                CONNECTION=0
                IO_ERROR=0
                WRONG_LENGTH=0
                WRONG_MAP=0
                WRONG_REDUCE=0
        File Input Format Counters
                Bytes Read=2250
        File Output Format Counters
                Bytes Written=77
```

图 40-7　运行 aqistat.jar

在 client 上查看统计结果，如图 40-8 所示。

```
[root@client ~]# hadoop fs -ls /aqioutput
[root@client ~]# hadoop fs -cat /aqioutput/part-r-00000
```

```
[root@client ~]# hadoop fs -ls /aqioutput
Found 2 items
-rw-r--r--   2 root supergroup          0 2016-12-14 04:49 /aqioutput/_SUCCESS
-rw-r--r--   2 root supergroup         77 2016-12-14 04:49 /aqioutput/part-r-00000
[root@client ~]# hadoop fs -cat /aqioutput/part-r-00000
严重污染        4
中度污染        14
优      52
良      66
轻度污染        36
重度污染        10
[root@client ~]#
```

图 40-8　查看各空气质量状况天数统计结果

实验四十一　综合实战：智能硬件大数据托管

41.1　实验目的

1. 了解智能硬件的开发过程；
2. 掌握基于万物云（www.wanwuyun.com）的智能硬件流程；
3. 熟悉智能硬件数据的上传、查询、分析。

41.2　实验要求

1. 本实验最好有硬件开发板，如果条件不允许可以参考本实验的模拟器方式。为了实验通用性我们的下一步骤均采用模拟器方式为同学们展示；
2. 试验结束时必须了解智能硬件的开发流程；
3. 试验结束时必须掌握智能硬件的数据上传、数据查询、数据分析。

41.3　实验原理

物联网智能硬件的开发需要硬件研发和软件研发相互配合，才能实现设备的互联互通，本实验主要从软件角度去了解智能硬件的开发。

硬件设备要想实现智能化，首先需要做到的是能将自己的传感器的数据传输到互联网中，通过收集传感器上传的数据，使用深度学习系统或其他分析系统判断设备的状态，然后通过互联网平台向设备下达基于数据分析之后的最优化的命令。实际的智能硬件研发中，硬件厂商往往只能研发硬件的数据采集设备，无法研发设备数据的托管，分析平台。

万物云是物联网设备和应用的数据托管平台。智能设备可使用多种协议轻松安全地向万物云提交所产生的设备数据，在服务平台上进行存储和处理，并通过数据应用编程接口向各种物联网应用提供可靠的跨平台的数据查询和调用服务。通过使用万物云平台所提供的各项服务，用户可以收集、处理和分析互联智能设备生成的数据，在物联网应用中方便地调用这些设备数据，而无须投资，安装和管理任何基础设施，不仅大大降低了项目开发的技术门槛，缩短开发周期，而且研发和营运成本也成倍降低。如图 41-1 所示。

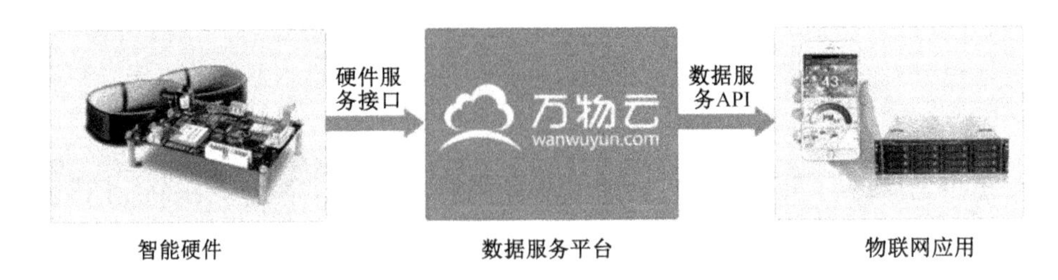

图 41-1　万物云使用流程

本实验主要使用 Java 去模拟硬件设备，然后使用模拟的代码完成硬件接口调用，向万物云平台存储数据，查询分析数据。实验中将会使用一个硬件接入的接口文档和一个数据查询的客户端，这个都是可以从万物云（www.wanwuyun.com）的官方网站下载的。万物云功能如图 41-2 所示。

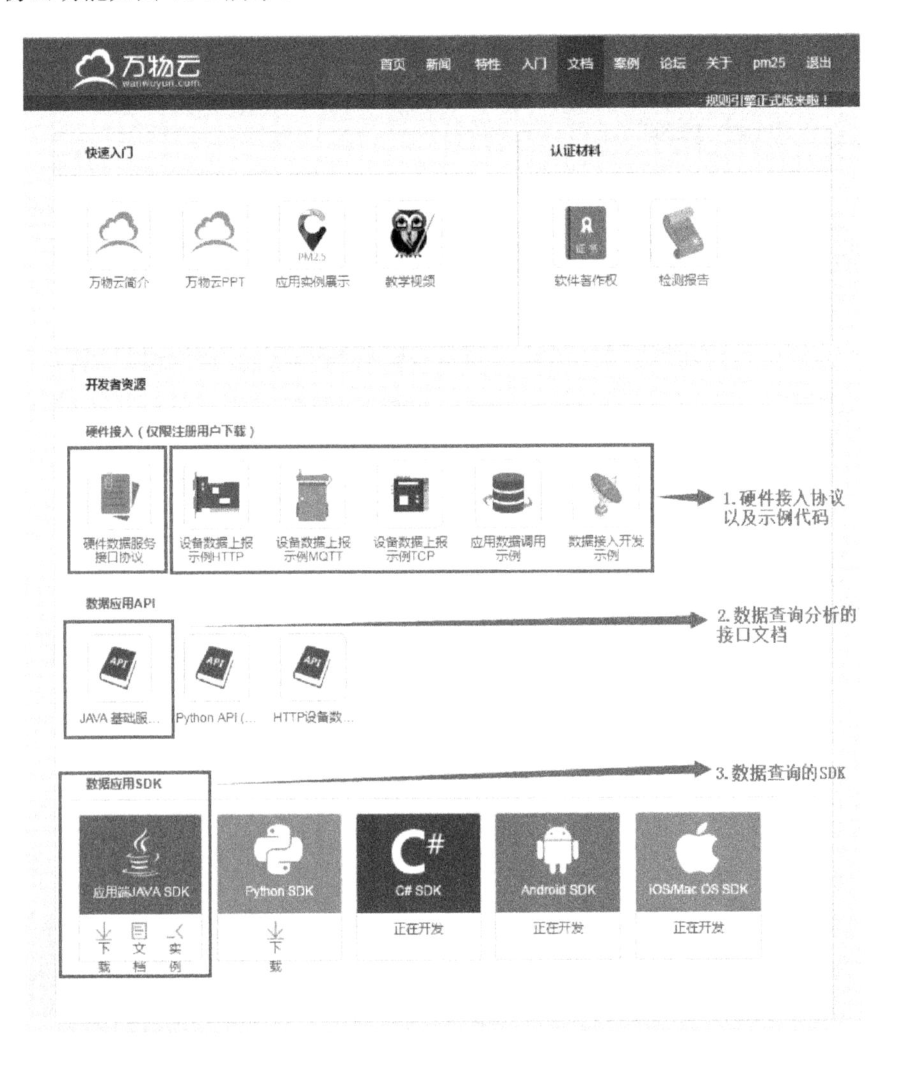

图 41-2　万物云功能

41.4　实验步骤

实验前先准备好实验原理中的文档和 jar 包，实验使用 Intellij IDEA 或者 Eclipse 为 IDE，先建立一个 Java 项目，用于模拟硬件客户端。本实验主要从以下介绍快速智能硬件开发步骤和流程：

（1）万物云平台相关注册；
（2）建表——用于存储智能硬件的数据；
（3）智能硬件接入平台；
（4）数据上传；
（5）数据查询；
（6）简单的数据分析。

41.4.1　万物云平台相关注册

首先，登录 www.wanwuyun.com，注册一个用户，如图 41-3 所示。

图 41-3　注册界面

然后使用注册的账号登录万物云平台，建立智能硬件 App，用于存储智能硬件设备数据，如图 41-4 所示。

图 41-4　应用中心界面

两个重要的信息：appid 应用 ID 和 accessid 安全管理中的安全码，在后续模拟代码中会用到。

41.4.2　建表——用于存储智能硬件的数据

单击应用，进入应用的数据表，添加智能硬件能够上报的数据列。本实验主要模拟一个带有 GPS 定位的温度和 PM2.5 测试的硬件环境设备。所以建表内容如下（建表操作可以调用 SDK 也可以直接在万物云平台上通过可视化界面建立，本实验采用可视化界面建立这样可以加速开发速度），如图 41-5 所示。

图 41-5　创建数据表界面

41.4.3　智能硬件接入平台

首先使用代码模拟一个硬件设备，然后完成智能设备接入平台，获取设备安全码，上报数据、查询数据、简单数据分析——以获取平均值为例。本实验代码结构如图 41-6 所示。

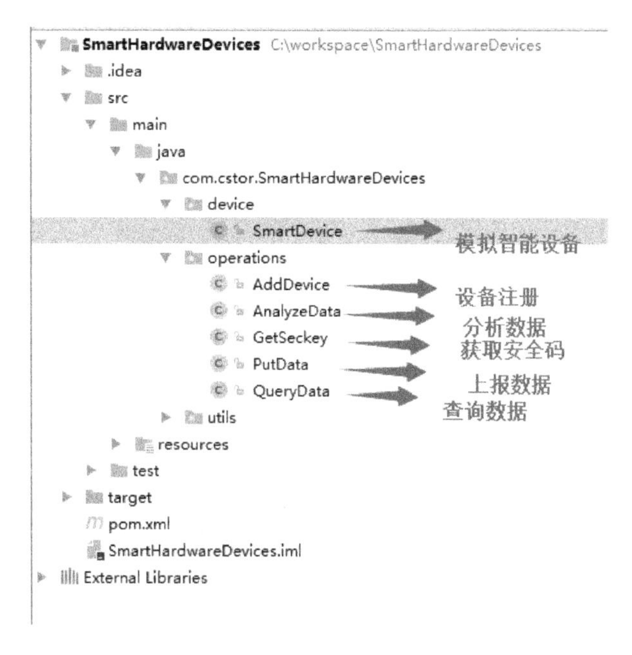

图 41-6　实验代码接口

万物云提供三种协议（TCP、MQTT 和 Http）用于接入智能硬件，本实验使用 TCP 用于演示，核心代码如下：

```
public class AddDevice extends ChannelHandlerAdapter {
    private static Logger logger = Logger.getLogger(AddDevice.class);
    //发送消息给服务器
    @Override
    public void channelActive(ChannelHandlerContext ctx) {
        logger.info("HelloClientIntHandler.channelActive");
        String msg = "[{\"user\":\"smart\",\"accessid\":\"C21HCNQXNDGXNTI4MTE2MTM2\",\"aid\":\
        "1528613677\",\"did\":\"st002\",\"sc\":\"A\"}]";

        System.out.println("设备注册信息: " + msg);
        ByteBuf encoded = ctx.alloc().buffer(4 * msg.length());
        encoded.writeBytes(msg.getBytes());
        ctx.writeAndFlush(encoded);
    }

    //接收 server 端的消息，并打印出来
    @Override
```

```
    public void channelRead(ChannelHandlerContext ctx, Object msg) throws Exception {
        logger.info("HelloClientIntHandler.channelRead");
        ByteBuf result = (ByteBuf) msg;
        byte[] result1 = new byte[result.readableBytes()];
        result.readBytes(result1);
        System.out.println("设备注册结果: " + new String(result1));
        result.release();
        ctx.close();
    }
}
```

41.4.4 数据上传

设备注册之后，可以通过万物云平台提供的接口进行数据上报，这个是核心的服务，主要是为智能硬件提供数据存储，其核心代码如下：

```
//发送消息给服务器
@Override
public void channelActive(ChannelHandlerContext ctx) {
    logger.info("HelloClientIntHandler.channelActive");
    String msg = "[{\"seckey\":\"uwmkpo2UPHbu1RN2lgAIBw\",\"row\":{\"dev_id\":\"st002\",
\"X\":\"12.456\",\"Y\":\"13.258\",\"T\":\"15\",\"PM25\":\"25\"}}]";
    System.out.println("模拟器采集的数据: " + msg);
    ByteBuf encoded = ctx.alloc().buffer(4 * msg.length());
    encoded.writeBytes(msg.getBytes());
    ctx.writeAndFlush(encoded);
}

//接收 server 端的消息，并打印出来
@Override
public void channelRead(ChannelHandlerContext ctx, Object msg) throws Exception {
    logger.info("HelloClientIntHandler.channelRead");
    ByteBuf result = (ByteBuf) msg;
    byte[] result1 = new byte[result.readableBytes()];
    result.readBytes(result1);
    System.out.println("数据上传结果: " + new String(result1));
    result.release();
    ctx.close();
}
```

41.4.5 数据查询

上传数据之后，可以使用万物云提供的 SDK 进行数据查询，希望同学们能够熟悉这个 SDK 的接口，能够熟练掉用其中的接口实现自己的功能，以下是查询数据的实验，其核心代码如下。

```
    public void getData() throws TException {
        //智能应用基本信息
        AppInfo app = new AppInfo();
        UserInfo user = new UserInfo();

        user.setUserName("smart");
        user.setAccessId("C21HCNQXNDGXNTI4MTE2MTM2");

        app.setAppId("1528613677");
        app.setUserInfo(user);
        //这个是智能应用数据存放的表
        String tableName = "DATA";
        //组装查询条件，这里主要组装了查询的起始时间和分页大小
        Conditions con = new Conditions();
        PageQueryCondition page = new PageQueryCondition();
        page.setPageSize(10);
        Date now = new Date();
        String st = Long.MAX_VALUE - now.getTime() + "";
        page.setStartPK(st);
        con.setPageQueryCondition(page);
        //调用万物云的客户端查询智能应用的数据
        ClientDevelopApi developApi = new ClientDevelopApiService();
        RePageObject ro = developApi.getTableRowsByNewPageCondition(app, "DATA", con, new )
ArrayList());
        System.out.println("查询结果是否正确返回: " + ro.isSuccess());
        System.out.println("数据结果如下: ");
        for (int i = 0; i < ro.getData().size(); i++) {
            Row row = (Row) ro.getData().get(i);
            System.out.println(row.getValues());
        }
    }
```

41.4.6 简单的数据分析

万物云能够提供数据托管服务，同时也能够提供数据分析的功能，本实验主要以最简单的设备数据均值分析为例，介绍如何使用万物云进行数据分析。

```
    public void getAvgData() throws TException {
        //设置智能应用的基本信息和用户的基本信息
        AppInfo app = new AppInfo();
        UserInfo user = new UserInfo();

        user.setUserName("smart");
        user.setAccessId("C21HCNQXNDGXNTI4MTE2MTM2");
```

```
        app.setAppId("1528613677");
        app.setUserInfo(user);
        //设置开始时间
        Date st = new Date();
        //设备号
        String devid = "st002";
        //设定均值的查询
        AvgQueryType avgQueryType = AvgQueryType.DAY;
        //调用 SDK 接口
        ClientDevelopApiService developApiService = new ClientDevelopApiService();
        ReJSONData ros = developApiService.queryDeviceDataAvgValues(app, devid, avgQueryType, st,
null,null);
        System.out.println("平均值查询是否正确:"+ros.isSuccess());
        System.out.println("时间:" + ros.getData().get("dateTime"));
        System.out.println("平均值:" + ros.getData().get("dataArray"));
    }
```

41.5 实验结果

本实验的每一步服务器都会返回相应的服务器代码，万物云官网是有这些代码解释的，如上报数据的时候返回码如图 41-7 所示。

2.2.2 返回结果：

Response Body JSON 数据格式：

```
{
    "code":"0"                    //返回编码
}
```

返回码：

```
0--成功
1--设备验证码错误
2--设备数据表不存在
3--IO 错误
4--入库数据格式不正确
6--版本不正确
```

* "[]"是协议的分割头和尾

** 红色字体部分为实例样本数据

*** 已建设备数据表表名：data， 其中 dev_id,pm1,pm2,...均为该应用设备数据表的相应数据字段名

图 41-7 实验结果代号解释图

其他每一步的代码都可以在网站上找到。

本实验做完之后是可以在万物云网站上看到自己建立的设备和设备上报的数据的，如图 41-8 所示。

图 41-8　数据查看页面

实验四十二　综合实战：贷款风险评估

42.1　实验目的

银行贷款员需要分析数据，以便搞清楚哪些贷款申请者是“安全的”，银行的“风险”是什么。这就需要构造一个模型或分类器来预测类标号，其预测结果可以为贷款员放贷提供相关依据。

本次实验通过提取贷款用户相关特征（年龄、工作、收入等），使用 Spark MLlib 构建风险评估模型，使用相关分类算法将用户分为不同的风险等级，此分类结果可作为银行放贷的参考依据。本次实验为方便演示，选用逻辑回归算法将用户风险等级分为两类：一类是高风险，另一类是低风险。有能力的同学可以尝试使用其他分类算法实现。

42.2　实验要求

1. 熟悉 Spark 程序开发流程：引入 jar 包→打包开发→打包→提交服务器运行；
2. 熟悉 Spark Mllib 中分类算法的使用流程；
3. 根据教程引导，对原始数据进行分类器模型训练。

42.3　实验原理

42.3.1　分类过程及评估指标

在使用分类算法进行数据分类时，均须经过学习与分类两个阶段。

1．学习阶段

（1）选定样本数据，将该数据集划分为训练样本与测试样本两部分（划分比例自定），训练样本与测试样本不能有重叠部分，否则会严重干扰性能评估。

（2）提取样本数据特征，在训练样本上执行选定的分类算法，生成分类器。

（3）在测试数据上执行分类器，生成测试报告。

（4）根据测试报告，将分类结果类别与真实类别相比较，计算相应的评估标准，评估分类器性能。如果性能不佳，则需要返回第二步，调整相关参数，重新执行形成新的分类器，直到性能评估达到预期要求。

2．分类阶段

（1）搜集新样本，并对新样本进行特征提取。

（2）使用在学习阶段生成的分类器，对样本数据进行分类。
（3）判别新样本的所属类别。

42.3.2　spark-submit 使用详解

将 jar 包提交服务器运行时，需要执行 spark-submit 相关命令才能运行，spark-submit 执行时命令格式如下。

```
spark-submit [options] <app jar | python file> [app options]
```

需要传入的参数说明如表 42-1 所示。

表 42-1　参数说明

参数名称	具体含义
master MASTER_URL	可以是 spark://host:port，mesos://host:port，yarn，yarn-cluster,yarn-client，local
deploy-mode DEPLOY_MODE	Driver 程序运行的地方，client 或者 cluster
class CLASS_NAME	主类名称，含包名
name NAME	Application 名称
jars JARS	Driver 依赖的第三方 jar 包
py-files PY_FILES	用逗号隔开的放置在 Python 应用程序 PYTHONPATH 上的.zip，.egg，.py 文件列表
files FILES	用逗号隔开的要放置在每个 executor 工作目录的文件列表
properties-file FILE	设置应用程序属性的文件路径，默认是 conf/Spark-defaults.conf
driver-memory MEM	Driver 程序使用内存大小
driver-library-path	Driver 程序的库路径
driver-class-path	Driver 程序的类路径
executor-memory MEM	executor 内存大小，默认 1G
driver-cores NUM	Driver 程序的使用 CPU 个数，仅限于 Spark Alone 模式
supervise	失败后是否重启 Driver，仅限于 Spark Alone 模式
total-executor-cores NUM	executor 使用的总核数，仅限于 Spark Alone、Spark on Mesos 模式
executor-cores NUM	每个 executor 使用的内核数，默认为 1，仅限于 Spark on Yarn 模式
queue QUEUE_NAME	提交应用程序给哪个 YARN 的队列，默认是 default 队列，仅限于 Spark on Yarn 模式
num-executors NUM	启动的 executor 数量，默认是 2 个，仅限于 Spark on Yarn 模式
archives ARCHIVES	仅限于 Spark on Yarn 模式

42.4　实验相关

42.4.1　实验环境

开发工具：IntelliJ IDEA；
操作系统：CentOS6.5；
编程语言：Scala 2.10.4；
相关软件：Hadoop2.6.0、Spark1.6.0。

42.4.2 实验数据

数据来源于：https://www.kaggle.com/，并作部分修改，源数据见附件 adult.csv。数据内容解释如下：

（1）risk-rating:0, 1；

（2）age: continuous；

（3）workclass: Private, Self-emp-not-inc, Self-emp-inc, Federal-gov, Local-gov, State-gov, Without-pay, Never-worked；

（4）fnlwgt: continuous；

（5）education: Bachelors, Some-college, 11th, HS-grad, Prof-school, Assoc-acdm, Assoc-voc, 9th, 7th-8th, 12th, Masters, 1st-4th, 10th, Doctorate, 5th-6th, Preschool；

（6）education-num: continuous；

（7）marital-status: Married-civ-spouse, Divorced, Never-married, Separated, Widowed, Married-spouse-absent, Married-AF-spouse；

（8）occupation: Tech-support, Craft-repair, Other-service, Sales, Exec-managerial, Prof-specialty, Handlers-cleaners, Machine-op-inspct, Adm-clerical, Farming-fishing, Transport-moving, Priv-house-serv, Protective-serv, Armed-Forces；

（9）relationship: Wife, Own-child, Husband, Not-in-family, Other-relative, Unmarried；

（10）race: White, Asian-Pac-Islander, Amer-Indian-Eskimo, Other, Black；

（11）sex: Female, Male；

（12）capital-gain: continuous；

（13）capital-loss: continuous；

（14）hours-per-week: continuous；

（15）native-country: United-States, Cambodia, England, Puerto-Rico, Canada, Germany, Outlying-US(Guam-USVI-etc), India, Japan, Greece, South, China, Cuba, Iran, Honduras, Philippines, Italy, Poland, Jamaica, Vietnam, Mexico, Portugal, Ireland, France, Dominican-Republic, Laos, Ecuador, Taiwan, Haiti, Columbia, Hungary, Guatemala, Nicaragua, Scotland, Thailand, Yugoslavia, El-Salvador, Trinadad&Tobago, Peru, Hong, Holand-Netherlands。

42.4.3 实验步骤

1．IDEA 配置

在 IntelliJ IDEA 中需要导入 Spark 开发包，Spark/lib 中的 jar 包能满足基本的开发需求，开发者可以在菜单：File→project stucture→Libraries 中设置，如图 42-1 所示。

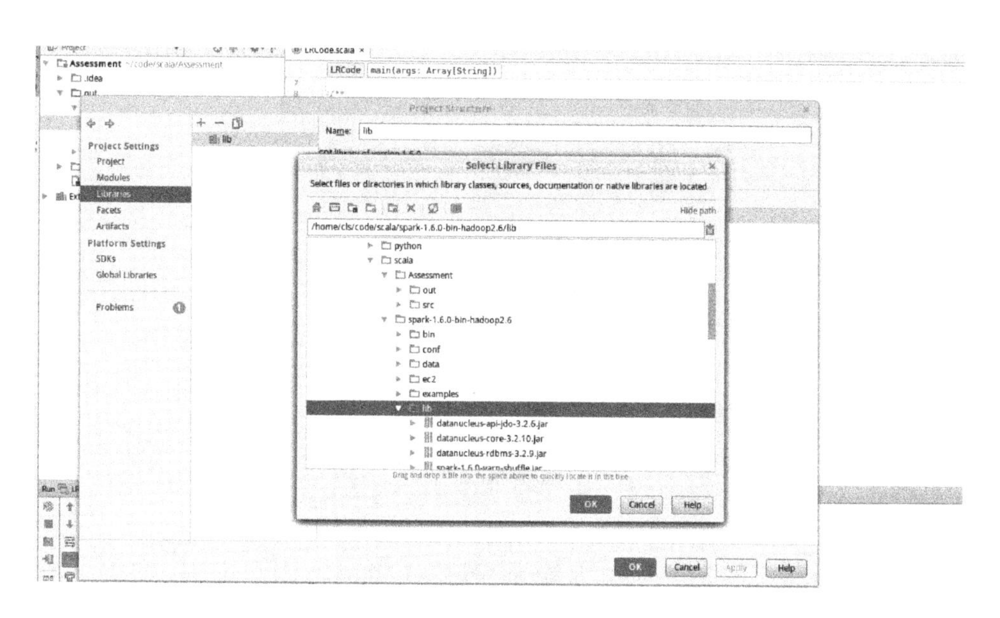

图 42-1　导入 Spark 包图

2．代码步骤

获取源数据。

```
val path = "hdfs://master:8020/input/adult.csv"
val rawData = sc.textFile(path)
```

简单的数据清洗。

```
/**
 * 取第一列为类标，其余列作为特征值
*/
val data = records.map{ point =>
val firstdata = point.map(_.replaceAll(" ",""))
  val replaceData=firstdata.map(_.replaceAll(","," "))
val temp = replaceData(0).split(" ")
  val label=temp(0).toInt
  val feature s = temp.slice(1,temp.size-1)
                .map(_.hashCode)
                .map(x => x.toDouble)
  LabeledPoint(label,Vectors.dense(features))
}
```

按照一定的比例将数据随机分为训练集和测试集。

这里需要程序开发者不断的调试比例，以达到预期的准确率，值得注意的是，不当的划分比例导致“欠拟合”或“过拟合”的情况产生。

```
val splits = data.randomSplit(Array(0.8,0.2),seed = 11L)
val traning = splits(0).cache()
val test = splits(1)
```

训练分类模型。

```
val model = new LogisticRegressionWithLBFGS().setNumClasses(2).run(traning)
```

预测测试样本的类别。

```
val predictionAndLabels = test.map{
case LabeledPoint(label,features) =>
val prediction = model.predict(features)
    (prediction,label)
}
```

计算并输出准确率。

```
val metrics = new BinaryClassificationMetrics(predictionAndLabels)
val auRoc = metrics.areaUnderROC()
println("Area under Roc =" + auRoc)
```

输出权重最大的前 10 个特征。

```
val weights = (1 to model.numFeatures) zip model.weights.toArray
println("Top 5 features:")
weights.sortBy(-_._2).take(5).foreach{case(k,w) =>
println("Feature " + k + " = " + w)
}
```

保存与加载模型。

```
val modelPath = "hdfs://master:8020/output/"
model.save(sc, modelPath)
val sameModel = LogisticRegressionModel.load(sc,modelPath)
```

3. 代码实例

```
import org.apache.spark.mllib.classification.LogisticRegressionModel
import org.apache.spark.mllib.classification.LogisticRegressionWithLBFGS
import org.apache.spark.mllib.evaluation.{BinaryClassificationMetrics, MulticlassMetrics}
import org.apache.spark.mllib.regression.LabeledPoint
import org.apache.spark.{SparkConf, SparkContext}
import org.apache.log4j.{Level, Logger}
import org.apache.spark.mllib.linalg.Vectors

object LRCode {
  def main(args:Array[String]): Unit = {
    val conf = new SparkConf()
                    .setAppName("Logisitic Test")
                    .setMaster("spark://master:7077")

    val sc = new SparkContext(conf)

    //屏蔽不必要的日志信息
```

```
Logger.getLogger("org.apache.spark").setLevel(Level.WARN)
Logger.getLogger("org.eclipse.jetty.server").setLevel(Level.OFF)

//使用 MLUtils 对象将 hdfs 中的数据读取到 RDD 中
val path = "hdfs://master:8020/input/adult.csv"
val rawData = sc.textFile(path)

val startTime = System.currentTimeMillis()
println("startTime:"+startTime)

//通过“\t”即按行对数据内容进行分割
val records = rawData.map(_.split("\t"))

/**
  * 取第一列为类标，其余列作为特征值
  */
val data = records.map{ point =>
  //去除集合中多余的空格
  val firstdata = point.map(_.replaceAll(" ",""))
  //用空格代替集合中的逗号
  val replaceData=firstdata.map(_.replaceAll(","," "))
  val temp = replaceData(0).split(" ")
  val label=temp(0).toInt
  val features = temp.slice(1,temp.size-1)
    .map(_.hashCode)
    .map(x => x.toDouble)
  LabeledPoint(label,Vectors.dense(features))

}

//按照 3:2 的比例将数据随机分为训练集和测试集
val splits = data.randomSplit(Array(0.8,0.2),seed = 11L)
val traning = splits(0).cache()
val test = splits(1)

//训练二元分类的 logistic 回归模型
val model = new LogisticRegressionWithLBFGS().setNumClasses(2).run(traning)

//预测测试样本的类别
val predictionAndLabels = test.map{
  case LabeledPoint(label,features) =>
```

```
            val prediction = model.predict(features)
            (prediction,label)
        }

        //输出模型在样本上的准确率
        val metrics = new BinaryClassificationMetrics(predictionAndLabels)
        val auRoc = metrics.areaUnderROC()
        //打印准确率
        println("Area under Roc =" + auRoc)

        //计算统计分类耗时
        val endTime = System.currentTimeMillis()
        println("endtime:"+endTime)
        val timeConsuming = endTime - startTime
        println("timeConsuming:"+timeConsuming)

        //输出逻辑回归权重最大的前 5 个特征
        val weights = (1 to model.numFeatures) zip model.weights.toArray
        println("Top 5 features:")
        weights.sortBy(-_._2).take(5).foreach{case(k,w) =>
          println("Feature " + k + " = " + w)
        }

        //保存训练好模型
        val modelPath = "hdfs://master:8020/output/"
        model.save(sc, modelPath)
    val sameModel = LogisticRegressionModel.load(sc,modelPath)

        //关闭程序
        sc.stop()
    }
}
```

4．服务器运行

1）编译器打包

（1）菜单：File→project stucture（也可以按快捷键 ctrl+alt+shift+s）。

（2）在弹窗最左侧选中 Artifacts→左数第二个区域点击“+”，选择 jar，然后选择 from modules with dependencies，然后会有配置窗口出现，配置完成后，勾选 Build On make （make 项目的时候会自动输出 jar）→保存设置，如图 42-2 所示。

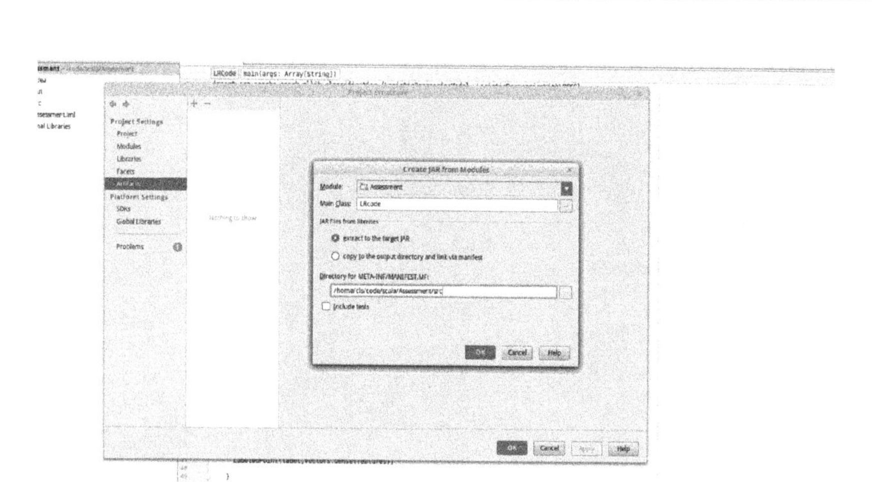

图 42-2　项目代码生成 jar 包图

（3）菜单：Build→make project。

（4）在项目目录下去找输出的 jar 包。

2）代码运行

（1）HDFS 中创建文件夹。

```
hadoop fs -mkdir /input
hadoop fs -mkdir /output
```

（2）将数据提交至 HDFS。

```
hadoop fs -copyFromLocal /root/data/42/adult.csv /input
```

（3）将 jar 包提交服务器，执行以下命令。

```
./bin/spark-submit --class LRCode --num-executors 3 --executor-memory 1g --executor-cores 3 /root/data/42/Assessment.jar
```

42.5　实验结果

由图 42-3 可知，该分类模型准确率约为 71.2%，耗时为 23579 毫秒，权重最大的前五个特征为第 5 个、6 个、11 个、12 个、13 个特征。

```
16/12/15 08:47:26 INFO optimize.LBFGS: Step Size: 0.02914
16/12/15 08:47:26 INFO optimize.LBFGS: Val and Grad Norm: 0.410850 (rel: 7.87e-08) 0.00364139
16/12/15 08:47:27 INFO optimize.LBFGS: Step Size: 1.000
16/12/15 08:47:27 INFO optimize.LBFGS: Val and Grad Norm: 0.410850 (rel: 2.33e-06) 0.00304491
16/12/15 08:47:27 INFO optimize.LBFGS: Step Size: 1.000
16/12/15 08:47:27 INFO optimize.LBFGS: Val and Grad Norm: 0.410847 (rel: 6.03e-06) 0.00153929
Area under Roc =0.7124486446831049
endtime:1481791655150
timeConsuming:23579
Top 5 features:
Feature 5 = 5.439472541058384E-4
Feature 13 = 5.7454444176084154E-5
Feature 12 = 8.020666775361346E-7
Feature 11 = 1.8109722375846685E-7
Feature 6 = 1.3149771490424918E-9
16/12/15 08:47:35 INFO Configuration.deprecation: mapred.tip.id is deprecated. Instead, use mapreduce.task.id
16/12/15 08:47:35 INFO Configuration.deprecation: mapred.task.id is deprecated. Instead, use mapreduce.task.attempt.id
16/12/15 08:47:35 INFO Configuration.deprecation: mapred.task.is.map is deprecated. Instead, use mapreduce.task.ismap
16/12/15 08:47:35 INFO Configuration.deprecation: mapred.task.partition is deprecated. Instead, use mapreduce.task.partition
16/12/15 08:47:35 INFO Configuration.deprecation: mapred.job.id is deprecated. Instead, use mapreduce.job.id
16/12/15 08:47:40 INFO hadoop.ParquetFileReader: Initiating action with parallelism: 5
SLF4J: Failed to load class "org.slf4j.impl.StaticLoggerBinder".
SLF4J: Defaulting to no-operation (NOP) logger implementation
SLF4J: See http://www.slf4j.org/codes.html#StaticLoggerBinder for further details.
16/12/15 08:47:41 INFO mapred.FileInputFormat: Total input paths to process : 1
```

图 42-3　实验结果输出图